Chapman & Hall/CRC Biostatistics Series

Dose Finding by the Continual Reassessment Method

Chapman & Hall/CRC Biostatistics Series

Mark Chang

DNA Microarrays and Related Genomics Techniques: Design, Analysis, and Interpretation of Experiments, David B. Allsion, Grier P. Page, T. Mark Beasley, and Jode W. Edwards

Dose Finding by the Continual Reassessment Method,
Ying Kuen Cheung

James E. De Muth

Elementary Bayesian Biostatistics,
Lemuel A. Moyé

Frailty Models in Survival Analysis,
Andreas Wienke

Handbook of Regression and Modeling: Applications for the Clinical and Pharmaceutical Industries, Daryl S. Paulson

Measures of Interobserver Agreement and Reliability, Second Edition,
Mohamed M. Shoukri

Medical Biostatistics, Second Edition,
A. Indrayan

Mikel Aickin

Meta-Analysis in Medicine and Health Policy, Dalene *Generalized Linear Models: A Bayesian Perspective,* Dipak K. Dey, Sujit K. Ghosh, and Bani K. Mallick

Karl E. Peace and

Monte Carlo Simulation for the Pharmaceutical Industry: Concepts, Algorithms, and Case Studies, Mark Chang

Multiple Testing Problems in Pharmaceutical Statistics, Alex Dmitrienko, Ajit C. Tamhane, and Frank Bretz

Anders Källén

Jay Herson

Sample Size Calculations in Clinical Research, Second Edition, Shein-Chung Chow, Jun Shao, and Hansheng Wang

Statistical Design and Analysis of Stability Studies, Shein-Chung Chow

Statistical Methods for Clinical Trials,
Mark X. Norleans

Statistics in Drug Research: Methodologies and Recent Developments, Shein-Chung Chow and Jun Shao

Karl E. Peace

Lemuel A. Moyé and

Statistics in the Pharmaceutical Industry, Third Edition, Ralph Buncher and Jia-Yeong Tsay

Translational Medicine: Strategies and Statistical Methods, Dennis Cosmatos and Shein-Chung Chow

Chapman & Hall/CRC Biostatistics Series

Dose Finding by the Continual Reassessment Method

Ying Kuen Cheung

Columbia University

New York, New York, USA

CRC Press
Taylor & Francis Group
Boca Raton London New York

CRC Press is an imprint of the
Taylor & Francis Group, an **informa** business

Preface

Despite its poor statistical properties, the 3+3 algorithm used dose finding method in phase I clinical trials today. How to realize the important role of dose finding in the drug devel is an increasing openness to "novel" methods proposed in the particular, the continual reassessment method (CRM) and it much attention in the medical community. To ride on this mome the status quo in the phase I practice, it is critical for us (s design a trial using the CRM in a timely and reproducible mann to writing a detailed exposition on the calibration of the CR who need to deal with dose finding in phase I trials while havin to attend to.

A natural approach to such a writing project is to write a *how-to* I started this book project in the summer of 2008, I had helped CRM trials (three of which are included as examples in this bo found some general patterns of how I calibrated the CRM param These patterns, characterized collectively as a trial-and worked well in the sense that they gave reasonable operating design. However, it was time-consuming (weeks of simulatio an intimate understanding of the CRM (I wrote a PhD dissertat realized that some automation and step-by-step guidelines would be crucial and appreciated if the CRM was to be used on a r a wide group of statisticians. Chapters 7–10 try to address t CRM design into a list of design parameters, each of which is t prescribed manner.

Despite my pragmatic approach, I hope this book is not only a c to provide a full coverage of the CRM. This book includes a com the CRM (Chapter 3) and elaborate properties of the CRM (Chap this book is based on my previous publications on the CRM, I ha material so as to present the CRM under a unified framework (Ch chapters serve as the theoretical foundation of the calibra in the later chapters. I also reflect on what *not* and when *not* to use the CRM (Chapter 13). From a practical viewpoint, these chapters are as important as, if not more important than, the *how-to* they avoid abuses and pitfalls in applying the CRM. I believe a wrong way or in the wrong trial is no better, or arguably wors to the 3+3 algorithm. The time-to-event aspect of the toxici

recurring concern in my previous CRM trials, and so is includ
the CRM (Chapter 11). All in all, while this is not intended to
inclusion of materials is based on their practical relevance.

This book does not cover dose finding in all possible clinical
it has a singular focus on the simplest and the most common pha
where the study endpoint is defined as a binary outcome and the
to come from a homogeneous population. I make no mention of th
multiple toxicity and the gradation of severe toxicities. T
dosing is omitted. While some basic ideas of dose finding usin
toxicity are outlined in Chapter 13, the discussion is brief
to this fast-growing area. All these are important topics in
interested. Their omission, however, is mainly due to my limited *practical*
in dealing with these "nonstandard" situations in real dose
with these issues simply from a *methodological* and *theoretical*
fit my intent of writing a practical book (although I think suc
in its own right and hope someone more qualified than I will del
a word or two to add from a methodological and theoretical vie
already alluded to in the book's final chapter (Section 14.4,
a complete theoretical framework is crucial for these nonst
successfully translated into actual practice. In this book
pathological behaviors (e.g., incoherence and rigidity) o
for the simplest setting; it is reasonable to infer that thes
for methods more complex than the CRM for the more complicate
Solid theoretical investigation will help us navigate the p
the theoretical framework developed in this book for the sim
useful when extended to the complicated settings. Second, a
I think stochastic approximation offers partial solutions
to many of these nonstandard dose finding settings. This is wh
with a chapter where I try to connect and compare the CRM with t
approximation literature.

The last points I just made give a hint about my methodologica
interests. I hope that this book will in some way simulate res
general dose finding methods, despite its practical nature.
CRM and the dose finding criteria at a rigorous level, and to co
as comprehensively as possible, I also hope this book can ser
for those interested in doing research in this area. I taught
experimentation at Columbia University from an early unpub
book. This final manuscript is, in turn, adapted from the cour
for use in a course on sequential experimentation or clinical trials.

There are several statistics books on dose finding. The two mo
are the edited volumes by Chevret [26] and by Ting [105]. Both
dose finding methods and are good introductions to the dose fin
comparison, this book is a single-authored work on a specific
which I think is necessary if we are to get down to the nuts and b

By writing a book on the CRM, I do not imply that it is the best me

In fact, for the dose finding objective considered here, it is
method that is best or optimal in a *uniform*
best under certain scenarios according to some criterion, t
a different criterion. There have been numerous proposals i
These proposals can be good alternatives against the 3+3 alg
are calibrated properly. And, the CRM is one of these methods
CRM has been worked out and discussed in the statistical and m
extensively that I believe we are getting close to translati
This book hopefully will be a catalyst in this translational process.

I owe a debt of gratitude to Tom Cook, Bin Cheng, and an anonymo
who have been generous with their time and given detailed com
versions of the book. I am grateful for Jimmy Duong for his hel
package 'dfcrm' (a companion software with this book). I wou
Rick Chappell who introduced me to the CRM and clinical trial
at University of Wisconsin–Madison. This book would not be p
mentoring. Finally, my most heartfelt thanks go to my wife, A
and enthusiasm during this writing process.

Contents

III CRM and Beyond

Part I

Fundamentals

Chapter 1

Introduction

The clinical development of a new drug or a new treatment proc
phases of testing in human subjects. Phase I trials are small
safety and identify a safe dose range of the treatment. Once a
sen, its therapeutic efficacy will be examined in a phase II tr
shown promising in phase II trials will be moved to multi-ins
cal trials for randomized comparison to standard treatment
entire process is to translate promising discoveries in the
procedures that can be used in the general clinical settings
trials, however, may give an oversimplified picture of the ac
process. Often, several phase I-II trial sequels of a drug, p
ations in the treatment schedule and patient populations, a
III trial is warranted. This process is necessarily iterati
phase I-II-III paradigm appears to suggest. In addition, th
universal across disciplines, and may include finer divisio
IIA, and IIB. The recent trend to combine phases of trials, th
phase I/II trials and seamless phase II/III trials, renders
development process.

This having been said, the phase I-II-III paradigm provides
work for in-depth study of statistical methods. The subject
finding using the continual reassessment method (CRM). The C
the first model-based designs for phase I cancer trials in whi
study endpoint. The role of toxicity in early-phase cancer t
ject for discussion in the medical literature [93, 85]. In pa
toxicity serves as evidence that the drug has reached a level
to the cancer cells but also to a patient's normal organs. In o
dose is expected to cause a significant amount of severe but re
the cancer patient population. Therefore, a primary goal of
identify the so-called maximum tolerated dose (MTD). For ot
acute stroke and HIV, identifying the MTD is also a primary ob
safety studies (usually called phase IB trials). In additio
in phase II proof-of-concept trials where the goal is to iden
demonstrated biological activity. This objective is usual
mation of the minimum effective dose (MED) [106, 27]. From a s

the MTD in safety studies and the MED in efficacy studies can be analogous way. Therefore, this book is relevant to the desig finding trials. Under the modernized paradigm, the dose findi here also address the design issues in the combined phase I/I the safety and the efficacy endpoints are considered as co-pr

Another advantage of dividing the drug development process by doing so, we can set a clear and manageable benchmark to ach study. This entails a clearly defined set of study endpoints a objective. Since clinical trials are conducted in human sub to be reached within certain ethical constraints. In partic studies, randomization is not entirely appropriate becaus excessively high and toxic doses without sufficiently testi would also argue randomization exposes subjects to low and i although this aspect is apparently not as alarming.)

We illustrate these points using a bortezomib dose finding tr is a proteasome inhibitor with proven activity in lymphoma. was given in combination with the standard chemotherapy as a for patients with diffuse large B cell or mantle cell non-Hod patient would receive up to six 21-day cycles of the treatmen describes the five dose schedules of bortezomib tested in the concerns related to bortezomib were neuropathy, low platel non-neurologic or non-hematologic toxicity. Toxicity was National Cancer Institute Common Terminology Criteria for with grade 3 or higher defined as *dose limiting* vere but can be resolved by symptomatic treatment, whereas a irreversible; toxic death due to the treatment is invariabl primary endpoint of each patient was the indicator of whethe icity (DLT) was experienced at any time during the six cycles trial was to determine the MTD, defined as a dose associated wi Table 1.1 gives the number of patients and the number of DLTs p zomib trial. The data show strong evidence that the highest d we pool the observations in dose levels 4 and 5 by assuming an i toxicity relationship; based on 1 DLT out of 16 patients, we o upper bound of 0.26 for the DLT probability.

Table 1.1 The bortezomib trial [62]: dose schedules of bortezomib, sample size (n) number of DLT (z) at each dose

Level	Dose and schedule within cycle	n	z
1	0.7 mg/m^2 on day 1 of each cycle	0	0
2	0.7 mg/m^2 on days 1 and 8 of each cycle	0	0
3	0.7 mg/m^2 on days 1 and 4 of each cycle	4	0
4	1.0 mg/m^2 on days 1 and 4 of each cycle	9	1
5	1.3 mg/m^2 on days 1 and 4 of each cycle	7	0

While simple analyses are usually adequate to address the pr questions in a phase I study, the summary statistics in Table were collected. Figure 1.1 shows the dose assignments of the order. The trial started at level 3, a dose schedule that the i safe to treat patients. Escalation to the next higher dose oc had been followed for several weeks without signs of toxicit took place after three following patients. Shortly after th trial at the highest dose, patient 7 at dose level 4 experienc a deescalation for the ninth patient. Subsequent patients w fashion, allowing months to pass before reescalating to the feature of this dose assignment scheme is its outcome adapti the bortezomib trial, the dose assignments were made in acco event continual reassessment method (TITE-CRM), an extens discussed in Chapter 11. For ethical reasons, most dose findi in an outcome-adaptive manner, so that the dose assignment o depends on those of the previous patients. As such, the focus *sign* (as opposed to analysis) of a dose finding study using the CRM a

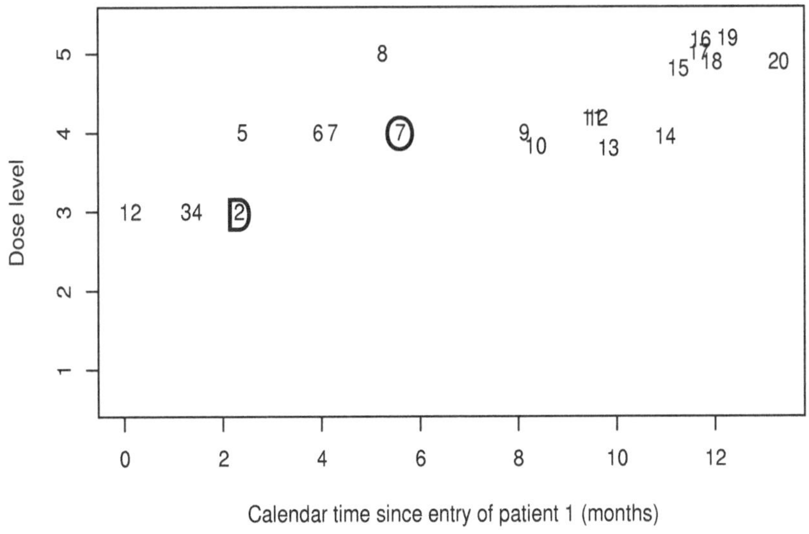

Figure 1.1 Dose assignments in the bortezomib trial. Each nu unmarked number represents the patient's entry time; a numb time when a DLT occurs, and "D" indicates the time of dropout. numbers are jittered for clarification.

This book is organized into three parts. Part I (Chapters 2–6 ground and introductory material of the CRM. Specifically, C clinical background, outlines the problem of dose finding in real trial examples, and reviews the dose finding literature basic approach of the CRM and presents its major modification be developed along with a description of an R package 'dfcrm' a unified framework for dose–toxicity models used in the CRM. respectively, discuss the theoretical and empirical prope jective of Part I is for the readers to develop a basic underst be able to implement the method using a simple R code. Readers basic CRM methodology are also encouraged to review the mate reorganized and presented in a unified framework in this book.

Part II (Chapters 7–10) details the calibration process of t notation and the theory introduced in Part I. Chapter 7 intro parameters involved in the CRM, and classifies them into two categories: *parameters* and *model parameters* techniques of the model parameters: the initial guesses of t (Chapter 8), the prior distribution of the model parameter (design of a two-stage CRM (Chapter 10). The objective of Part to develop the ability to design a "good" CRM trial within a reasonable timeline.

Part III (Chapters 11–14) contains a variety of advanced top CRM. Chapter 11 presents the TITE-CRM to deal with situation ity outcome is defined with respect to a nontrivial duration. review of CRM using multiparameter models. Chapter 13 consi the CRM is an inappropriate design, and puts forward some alt connects the CRM and modern dose finding trials to the large li approximation. The objective of Part III is to stimulate fur and general dose finding methodology.

The materials in this book are presented at a level that requi and some basic calculus concepts. Sections marked with "†" i contain technical details that may be skipped without affec other chapters. Exposition in the book will be supplemented usage of R functions in the 'dfcrm' package. While some basic enhance the reading experience, proficiency in R is not requi can find out more information about R from the Comprehensive R (CRAN) [83] at

`http://www.r-project.org.`

Chapter 2

Dose Finding in Clinical Trials

2.1 The Maximum Tolerated Dose

The primary objective of phase I trials of a new anticancer dr
toxic side effects of the drug and to recommend a dose for the s
II trials. This recommended dose is typically the maximum te
exceed an acceptable level of toxicity, the so-called *maximum tolerated dose*
Traditional chemotherapy takes the cytotoxic therapeutic
toxicity may be viewed as a surrogate for anti-tumor activit
in a sense a desirable endpoint, so the trial objective is to fi
with a given level of toxicity probability. Also in this sens
optimal in the absence of information about efficacy and clinical response.

Definition 2.1 (MTD—surrogacy perspective).
denote the toxicity probability associated with dose level k for $k = 1, \ldots,$
MTD from a surrogate-for-efficacy perspective, denoted by ν
level with toxicity probability closest to a prespecified target probability θ
$\nu \equiv \arg\min_k |p_k - \theta|$.

Since the late 1980s, most dose finding designs have been prop
address dose finding of cytotoxic drugs in patients with soli
of malignancies. The bortezomib trial in lymphoma patients
is one such example. As a result, the phase I method literatur
surrogacy definition of MTD. See Exercise 2.2 for an alternat
from a surrogacy perspective.

For noncytotoxic target anticancer agents and treatments f
as acute ischemic stroke, toxicity does not play a therapeut
the primary concern in the early drug development phase. For
useful to define an upper safety limit of the dose range for fur
Furthermore, under the assumption that efficacy of a drug inc
is merit in pushing the dose as high as safety will permit. In v
off between safety and efficacy, one may seek to maximize the d
to patients subject to toxicity constraints. This may lead t
recommendation for the next study phase:

Definition 2.2 (MTD—trade-off perspective).

denote the toxicity probability associated with dose level k for $k = 1, \ldots,$
MTD from a trade-off-for-efficacy perspective, denoted by γ
dose level with toxicity probability no greater than a prespecified threshold θ
is, $\gamma \equiv \max\{k : p_k \leq \theta\}$.

Both the surrogacy and the trade-off perspectives define the
a target toxicity rate θ
problem. However, from a statistical viewpoint, it is gener
than γ (if ν is well defined) by making use of monotonicity of the dose–tox
curve. Therefore, it is in some cases pragmatic to take ν
a trial, even though the toxicity endpoint is generally not a
noncytotoxic drugs. In practice, it is important to discern
for a given clinical setting, and the degree of tolerance in t
Apparently, these decisions need to be made on a trial-by-trial basis.

Example 2.1 (acute ischemic stroke).
early after stroke in animal models, have demonstrated neur
in a dose-dependent manner, with the greatest effects at the
NeuSTART (Neuroprotection with Statin Therapy for Acute Re
development program aimed to translate preclinical resear
dose statins in stroke patients. In a phase IB dose finding stu
program [34], high-dose lovastatin was given to patients fo
lowed by a standard dose for 27 days. The primary safety conce
dose lovastatin included elevated liver enzyme; a toxicity
enzyme levels at any posttreatment time points exceeded a pr
There were five test doses in the NeuSTART and the trial object
dose with toxicity rate closest to 10%, that is, $\theta = 0.10$.

Example 2.2 (cancer prevention).
that is thought to block tumor promotion by inhibiting cell p
cell cycle arrest and apoptosis. The multiple mechanisms of
candidate agent for chemoprevention. On the other hand, the
toxic and causing mortality in female beagle dogs after an ov
generally involved the gastrointestinal (GI) system, prod
to the lining of the GI tract, with hemorrhage and necrosis ap
Poly E trial was conducted in women with a history of hormone r
breast cancer. Three different doses of Poly E were administ
6 months. The study objective was to find the MTD defined as a dos
25% DLT during the six-month period, where a DLT here meant an
higher toxicity that would persist for at least one week or re
drug. As a secondary objective, this study included biologi
based biomarkers and mammography. Therefore, some subject
receive a placebo, to which the identified MTD would be compared.

Example 2.3 (early neuro-rehabilitation).

enhance recovery in stroke patients. On the other hand, prem
apy might cause neurologic worsening, cardiac complicatio
short term. Based on historical data, the adverse event rate
in the untreated patients during the first 4 days after stroke
Stroke Collaboration with Early Neurorehabilitation Ther
trial of early physical therapy (PT) in stroke patients, its
the largest PT dose that could be instituted without causing
of the 25% background rate. Table 2.1 displays the six PT regi
this example, a dose is composed of the timing and duration of therapy.

Table 2.1 Physical therapy (PT) regimen in ASCENT

Regimen	Minutes of PT on			Total PT dose
	Day 2	Day 3	Day 4	(minutes)
1	0	0	30	30
2	0	30	30	60
3	30	30	30	90
4	30	30	60	120
5	30	60	60	150
6	60	60	60	180

In the NeuSTART in Example 2.1, liver enzyme elevation is exp
3% of the stroke population, which consists mainly of the eld
toxicity endpoint is quite mild and reversible upon drug wit
higher than the background would be tolerated for the potent
and it would also seem reasonable to accept a dose above the ta
toxicity probability at this dose was close to θ
was chosen as the operative MTD objective for the trial, even
surrogacy perspective is far from reality: Liver enzyme ele
surrogate for any clinical benefits in stroke patients. Simi
for using ν as the objective in the Poly E trial; cf. Exercise 2.1.

On the other hand, in the ASCENT trial in Example 2.3, functio
to early physical therapy may not warrant an elevated advers
target rate θ should be set at the background 25%. In this situation, Defini
appears more appropriate than Definition 2.1. For one thing,
the dose–toxicity probability is about 25% for doses below a
objective ν is not uniquely defined. Figure 2.1c displays the plausible s
dose–toxicity curve in the ASCENT trial, under which γ
ally, the use of ν requires a strictly increasing dose–toxicity relationshi
MTD, whereas γ always exists regardless of the shape of the dose–toxicity c

Figure 2.1 also shows plausible dose–toxicity curves for th
the Poly E trial, where the MTD is defined with respect to a 25% t
In the bortezomib trial, since untreated lymphoma patients
a grade 3 neuropathy, the probability (y-intercept) in Figu

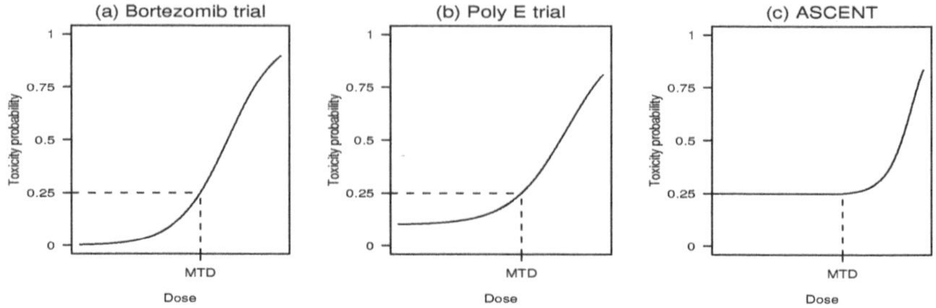

Figure 2.1 Plausible shapes of dose–toxicity curve on a conc
in three studies and the corresponding MTD, defined as a dose a
toxicity rate. Definition 2.1 is used to define MTD in (a) and (b), and Definition 2.2 in (c).

dose decreases. In contrast, because low-grade GI toxiciti
subjects with a history of cancer in the Poly E trial, there is
in Figure 2.1b. However, it is generally believed that the do
strictly increasing around the 25th percentile in both trials, and thus ν

It is easy to verify that $\nu = \gamma$
strictly increasing function of dose. Thus, if a continuum o
for the trial, it will be a practical choice to use ν
estimating ν is generally easier than estimating γ
practice allow only a discrete number of test doses; in which
may yield a slightly more aggressive recommendation than De
$\nu \geq \gamma$. Therefore, the choice between ν and γ
to be (slightly) more aggressive than the target θ
the nature of treatment, the severity of the disease, and the
endpoint.

As alluded to earlier, the CRM is motivated by applications i
a surrogacy view. Thus, the book will naturally focus on the estimation of ν
the CRM. However, Chapter 13 will explore situations in whic
applicable, and introduce an alternative approach that uses γ
as the MED in efficacy trials may be defined analogously to ν
applicable to dose finding in phase II efficacy trials. Howeve
on the MTD finding by the CRM according to the method's origina
the design strategy for the MED can be derived by analogy.

2.2 An Overview of Methodology

This section gives a brief overview of the development of the
since the late 1980s so as to put the CRM in a historical light.
of this book, the review will be cursory. Interested readers
in the edited volume by Chevret [26] and the article by Le Tourneau et al. [59].

The 3+3 algorithm. Traditionally, a 3+3 algorithm is used to dictate dose escal
and to approach the eventual recommended dose. The method st
low dose (e.g., one-tenth of LD10 in mice) and escalates afte
patients per dose; the recommended dose is defined as the larg
than two patients experiencing a predefined DLT during the fir
Table 2.2 describes the dose escalation rules of this algori
be slight variations from institution to institution. For e
preliminary information about efficacy of the drug, it is com
patients (usually 6 to 12) at the identified MTD.

Table 2.2 Escalation rules at a given dose with cumulative sample size n
of DLT in accordance with the 3+3 algorithm

n	z	Action
3	0	Escalate to the next higher dose
3	1	Treat three additional patients at the current dose
6	1	Escalate to the next higher dose
3 or 6	≥ 2	Stop escalation and terminal trial[a]

[a]*The MTD is estimated by the dose immediately below the terminating dose.*

The 3+3 algorithm has historically been the most widely used
Its main advantage is simplicity. Since the dose escalation
before a trial starts, the clinical investigators can make d
without help from a statistician.

However, the algorithm has two major shortcomings. First, d
dose and the conservative escalation scheme, the 3+3 algori
patients at low and inefficacious doses. Since phase I cancer
patients as subjects, as opposed to healthy volunteers, the
subjects at a therapeutic dose that is likely higher than the
the algorithm is discordant with the therapeutic intent of t
the 3+3 algorithm has no statistical justification. There is
method to stop escalation at any given percentile, and thus t
recommended MTD depends arbitrarily on the underlying dose
the number of test doses [97]. This deficiency is due to the fac
correspondence between the method and any quantitative defi
the 3+3 algorithm does not involve an explicit choice of the target θ
selection in the early phase will likely be carried over to it
phases, the use of the 3+3 algorithm will have lingering finan
adverse scientific consequences. And thus, the simplicity o
justify its widespread application.

Stochastic approximation.
can be traced back to the 1960s [93], a formal statistical for
appeared at a much later time. Among the earliest discussion
in 1984 proposed the use of stochastic approximation [86] in

procedure assigns dose sequentially by the recursion

$$x_{i+1} = x_i - \frac{1}{ib}(Y_i - \theta)$$

for some prespecified constant $b > 0$, where x_i is the dose assigned to patient i
Y_i is the toxicity indicator. The stochastic approximation is
in that it does not assume any parametric structure on the dos
Let $\pi(x) = \Pr(Y_i = 1 \mid x)$ denote the probability of toxicity at dose x
assumptions of $\pi(x)$, the dose sequence $\{x_i\}$
probability 1 to a dose x^* such that $\pi(x^*) = \theta$.

However, the use of recursion (2.1) implicitly assumes a con
is available for testing in the trial. This is not always feas
situations such as the bortezomib trial, there may be no natu
rather, "dose" is composed of drug dosage and treatment sche
other difficulties from a statistical viewpoint. First, the
been shown to be inferior to model-based methods such as the m
recursion for binary data [112, 50]. Second, the choice of the constant b
a large impact on the performance of the procedure. As a conse
approximation has seldom been used in dose finding trials. We
Chapter 14.

Up-and-down designs.
up-and-down schemes originally described by Wetherill [11
up-and-down design, which Storer called design D, is to enro
three; then, escalate dose for the next group if there is no to
group, deescalate if there is more than one toxicity in the gr
level if one of three patients has toxicity. By Markov chain r
D can be shown to sample around a dose that causes toxicity wit
$\theta = 0.33$. The group size and the decision rules in the up-and-down s
apparently be modified to accommodate other target θ
$\theta = 0.20$, one will intuitively enroll patients in groups of five. Al
schemes can be applied in stages. Storer, in particular, sug
of one initially in the trial and switching to design D upon th
The idea here is to move the trial quickly through the low dose
many patients at low and inefficacious doses.

Durham et al. [32] proposed a randomized version of the up-an
any target at or below the 50th percentile, that is, $\theta \leq 0.$
the dose for the next patient if the current patient has toxic
to a biased coin with probability $\theta/(1-\theta)$
a *biased coin design.*

In a trial that uses design D or the biased coin design for dose
the enrollment is complete, we may naturally estimate the do
logistic regression [98] or isotonic regression [99], and e
100θth percentile of the fitted curve. The consistency condition
the MTD hold, because the design points $\{x_i\}$
walk sampling plans. That is, technically, it can be shown th

treated at each dose will grow indefinitely as sample size inc
shown by the properties of random walk that the asymptotic di
allocation has a mode near the target MTD. However, from a *design*
"memoryless" property of random walk may cause ethical diffi
up-and-down rules make dose decisions based only on the most
group of patients at the current dose, previously accrued da
will likely reescalate to a dose that appears toxic.

Model-based designs.
on a dose–toxicity model, which is being updated repeatedly
as data are accrued. The continual reassessment method (CRM
O'Quigley et al. [78] in 1990, is the first model-based design
finding literature. Several model-based designs proposed s
notion with the CRM. One example is the escalation with overd
design [4], which takes the continual reassessment notion b
with respect to an asymmetric loss function that places heav
dosing than underdosing; see (2.2) below. A list of model-ba
Section 2.3.

Most model-based methods take the myopic approach by which d
is optimized with respect to the next immediate patient with
patients. For example, the EWOC at each step minimizes the Ba
to the loss function:

$$x_{i+1} = \arg\min_{x} E_i \left\{ \alpha(\nu - x)^+ + (1-\alpha)(x-\nu)^+ \right\}$$

where $x^+ = \max(x, 0)$ and $E_i(\cdot)$
posterior distribution of the MTD ν, given the first i
is intended to control overdose, the loss function (2.2) sho
value of α, that is, $\alpha < 0$.
optimization approach that minimizes the global risk and pr
$\{x_1, x_2, \ldots, x_N\}$ sequentially so as to mininize

$$E\left\{ \sum_{i=1}^{N} \alpha(\nu - x_i)^+ + (1-\alpha)(x_i - \nu)^+ \right\},$$

where the expectation is taken with respect to the joint distribution of x_i
Such sequential optimization is implemented by backward in
namic programming which can be computationally intensive.
presents a new direction for model-based design and warrants further research.

The model-based approach facilitates borrowing strength f
across doses through the parametric assumptions on the dose
especially important in early-phase dose finding trials whe
and informational content is low [41]. In addition, since th
a trial at a dose higher than the lowest level, the in-trial al
around the target dose.

Many model-based designs take a Bayesian approach. They upd

the uncertainty about the dose–toxicity curve with respect
given the interim data. This approach is by nature automated
computations can be efficiently programmed and reproduced.
automation is that a carefully *calibrated* dose–toxicity model can handle
contingencies in a manner that is coherent with the trial obj
may imagine the predicament arising with the 3+3 rule in a tri
two toxic outcomes among 7 patients at a dose; this contingen
overaccrual due to administrative delays, and is not uncommon in practice.

Several practical difficulties may hinder the use of a model-
there is skepticism among the clinicians because of the "bla
designs. Second, there is the perception that the success of
to the choice of the dose–toxicity model. Third, as a model-b
specialized computations, the clinical team will need to in
study statistician for interim dose decisions. Such intera
adding unnecessary burdens on both parties, in light of the f
algorithm requires minimal statistical inputs.

This book attempts to address the second difficulty with a foc
By theoretical and empirical arguments, we will see that the
does not depend on correctness of the model specification. Th
addresses the third difficulty by illustrating how a CRM desi
implemented in practice. The ultimate goal is to alleviate t
during the planning stage and the conduct of the trial. This e
availability of software. This book focuses on the R package
and its major variants. Software for some model-based desig
public access; see Table 2.3.

Table 2.3 Some software links for model-based approaches

Description		
Escalation with overdose control (EWOC)[a]	2.2	[4]
Late-onset toxicity monitor using predicted risks[b]	11.4.2	[8]
Modified CRM on a continuous dosage range[c]	12.4.4	[81]
Phase I/II dose finding based on efficacy and toxicity[b]	13.3	[102]

[a] *http://www.sph.emory.edu/BRI-WCI/ewoc.html*
[b] *http://biostatistics.mdanderson.org/softwaredownload*
[c] *http://www.cancerbiostats.onc.jhmi.edu/software.cfm*

Algorithm-based designs.
model-based designs, there seems to be a renewed interest in
designs since the late 1990s. Generally, an algorithm-base
of escalation rules for any given dose without regard to the o
As a result of the independence among observations across do
tabulated and made accessible to the clinical investigators before a trial starts.

The 3+3 algorithm is the most prominent example of algorithm
Efforts have recently been made to extend this traditional m

well-defined statistical properties. In particular, Cheun
as a multiple testing problem and introduces a class of stepw
operate in a manner similar to the 3+3 algorithm. This approa
peal because clinicians are familiar with the 3+3 algorithm
details of the stepwise procedures.

Ji et al. [49] propose a class of up-and-down designs that mak
based on the posterior toxicity probability intervals. Spe
of the toxicity probability p_k at dose k
ified constants $K_1, K_2 > 0$,

$$\Theta_{k,E} = \{p_k - \theta < -K_1\sigma_k\}$$

$$\Theta_{k,S} = \{-K_1\sigma_k \leq p_k - \theta \leq K_2\sigma_k\}$$

$$\Theta_{k,D} = \{p_k - \theta > K_2\sigma_k\}$$

so that $\Theta_{k,E} \cup \Theta_{k,S} \cup \Theta_{k,D} = [0,1]$, where σ_k
Escalation from a current dose, say dose k
of $\Theta_{k,E}$ is largest among the three sets. Similarly, deescalation occurs if $\Theta_{k,D}$
most probable event according to the posterior distribution of p_k
dose will remain at level k
random walk up-and-down is that the posterior interval uses
to a dose to make a dose decision and avoids the memoryless pro
walk design.

2.3 Bibliographic Notes

The practice and design of phase I trials are discussed in the
by Schneiderman [93], Carbone et al. [13], Ratain et al. [85,
Benjamin [55]. Korn [53] discusses the relevance of MTD in no
cancer treatments. Discussion of dose finding strategies in
is relatively sporadic. Fisher et al. [38] present some phas
the context of acute stroke MRI trials, and make a case agains
algorithm in the phase I safety studies. Cheung et al. [23] ma
for early-phase trials in patients with amyotrophic latera
design of NeuSTART is reported in Elkind et al. [34] as a case study.

Robbins and Monro [86] introduce the first stochastic approx
which has been studied extensively and has motivated a large
modifications. (See for example Sacks [92], Venter [107], La
and Wu [112, 113].) In the more recent literature, Lai [56] gi
the advances of stochastic approximation. Cheung [19] draw
this area and modern dose finding methods.

Lin and Shih [64] study the operating characteristics of a class of $A+B$
that include the 3+3 algorithm as a special case. The theoret
biased coin design are established by Durham and colleagues [29, 30, 31].

Several model-based designs have been proposed since the 19
the Bayesian decision-theoretic design [111], the logisti

and the Bayesian c
and will be reviewed in the next chapter. Chapter 12 will comm
dose finding designs: the curve-free method [40] and the isot
the EWOC, Zacks et al. [117] prove that the method is Bayesian
optimal, and consistent under the assumption that the speci
is correct. These Bayesian criteria are introduced in the pr
and Zacks [33].

2.4 Exercises and Further Results

Exercise 2.1. Discuss the MTD objective (surrogacy versus trade-off pers
for cancer prevention in the context of the Poly E trial.

Exercise 2.2. Definition 2.1 formulates dose finding as estimating a percen
dose–toxicity curve. Another possible alternative to defin
surrogacy perspective is

$$\nu' = \arg\min_k |\pi^{-1}(p_k) - \pi^{-1}(\theta)|.$$

Show that $|\nu - \nu'| \leq$
level. Discuss why ν' may not be applicable for the bortezomib trial.

Exercise 2.3. By computer simulations, generate the outcomes of a trial us
3+3 algorithm with $K = 5$ doses and true toxicity probabilities $p_1 = 0.02$, $p_2 = 0$
$p_3 = 0.10$, $p_4 = 0.25$, and $p_5 = 0$.
simulations 1000 times, and record the distribution of the r
1000 simulated trials.

Chapter 3

3.1 Introduction

In this book, we consider two types of dose finding strategies reassessment method (CRM): a one-stage design that necessi CRM (Section 3.2) and a two-stage design (Section 3.3). This basic CRM approach and introduces the necessary notation fo in the later chapters. Section 3.4 presents some simulation illustrate how the method may operate in practice. Section 3 modifications of the CRM. Section 3.6 gives key references in

3.2 One-Stage Bayesian CRM

3.2.1 General Setting and Notation

Consider a trial with K test doses with numerical labels $d_1, \ldots, d_K$ trial, patients are enrolled in small groups of size $m \geq 1$. Let $x_i \in \{d_1, \ldots, d_K\}$ the dose assigned to the i receive the same dose. In what follows, we first consider a *fully sequential* plan (i.e., $m = 1$), where we observe a binary toxicity outcome Y_i from the i and postulate Y_i as a Bernoulli variable with toxicity probability $\pi(x_i)$, where $\pi($ a monotone increasing function in x is, $m > 1$, in Section 3.5.2.

In accordance with Definition 2.1, the trial objective is to i $\nu \in \{1, \ldots, K\}$ that is associated with a toxicity probability θ.

3.2.2 Dose–Toxicity Model

The CRM assumes a dose–toxicity model $F(x, \beta)$ curve $\pi(x)$ is postulated to be $F(x, \beta_0)$ for some true parameter value β_0 the CRM does not require F to be a correct model for π, and β_0 as a "least false" value. Briefly, we require $F(x, \beta)$ x, in addition to some regularity conditions. Details of the a in Chapters 4 and 5. The two most commonly used models in the CR the empiric function

$$F(x, \beta) = x^{\beta} \text{ for } 0 < x <$$

and a one-parameter logistic function

$$F(x,\beta) = \frac{\exp(a_0+\beta x)}{1+\exp(a_0+\beta x)} \quad \text{for } -\infty < x < \infty$$

where the intercept a_0
the hyperbolic tangent function [78, 67]

$$F(x,\beta) = \left(\frac{\tanh x+1}{2}\right)^{\beta} \quad \text{for } -\infty < x < \infty.$$

To ensure an increasing dose–toxicity relationship, the parameter β
is restricted to taking on positive values. The positivity c
difficulty in estimation, especially when the sample size is
to consider the following parameterization:

$$\text{empiric: } F(x,\beta) = x^{\exp(\beta)} \quad \text{for } 0 < x <$$

$$\text{logistic: } F(x,\beta) = \frac{\exp\{a_0+\exp(\beta)x\}}{1+\exp\{a_0+\exp(\beta)x\}} \quad \text{for } -\infty < x < \infty$$

$$\text{hyperbolic tangent: } F(x,\beta) = \left(\frac{\tanh x+1}{2}\right)^{\exp(\beta)} \quad \text{for } -\infty < x < \infty$$

under which the parameter β is free to take on any real values while $F(x,\beta)$
increasing.

The original formulation of the CRM uses a Bayesian approach
model parameter β is assumed random and follows a prior distribution $G(\beta)$
will focus on the normal prior distribution, that is,

$$\beta \sim N(\hat{\beta}_0, \sigma_\beta^2)$$

where $\hat{\beta}_0$ and σ_β^2 are, respectively, the prior mean and variance.

3.2.3 *Dose Labels*

An important point about the CRM is that the numerical dose labels $d_1,\ldots,d_K$
not the actual doses administered, but rather are defined on a con
represents an ordering of the risks of toxicity. Consider th
the bortezomib trial (Table 1.1). The first three levels pres
dose 0.7 mg/m^2 with increasing frequency, whereas the next two levels appl
same frequency with increasing bortezomib dose. While it is
that the toxicity risk increases with each level, there is no
mg/m^2) in this application. Similarly, in the ASCENT trial (Examp
a composite of timing and duration of physical therapy given
is artificial to assume the dose–toxicity curve $\pi(x)$

dose range. Instead, one will have access only to a discrete s
As the CRM operates on a discrete set of dose levels, a physica
dose labels $d_1, \ldots, d_K$ is not required, as long as they constitute a *strictly*
sequence.

In practice, to ensure monotonicity, the label d_k
the initial guess of toxicity probability p_{0k} for dose level k
model, that is, solving

$$p_{0k} = F(d_k, \hat{\beta}_0).$$

The set of initial guesses $\{p_{0k}\}$
is a *strictly* increasing sequence, that is, $p_{01} < p_{02} < \cdots < p_{0K}$.

Suppose there is a prior belief that dose level ν_0
initial guess

$$p_{0,\nu_0} = \theta.$$

Consider, for instance, a trial with $K = 5$ dose levels and a target probability $\theta =$
Suppose we use the logistic function (3.3) with $a_0 = 3$ and prior mean $\hat{\beta}_0 =$
we believe that $\nu_0 = 3$ is the prior MTD such that $p_{03} = 0.$

$$0.25 = \frac{\exp\{3 + \exp(0)d_3\}}{1 + \exp\{3 + \exp(0)d_3\}}$$

and obtain $d_3 = -4.10$. Suppose further that $p_{01} = 0.05, p_{02} = 0.12, p_{04} =$
and $p_{05} = 0.55$. Then in the same manner, we obtain $d_1 = -5.94, d_2 = -$
$d_4 = -3.41$, and $d_5 = -2.80$. Table 3.1 shows the dose labels obtained by
ward substitution (3.5) under various dose–toxicity functions using the same
ton. Note that the range of d_k
However, it is always true that $d_1 < \cdots < d_K$.

Table 3.1 Dose labels via backward substitution under four CRM models for $K =$
$p_{01} = 0.05, p_{02} = 0.12, p_{03} = 0.25, p_{04} = 0.40, p_{05} = 0.55$, and prior mean $\hat{\beta}_0 = 0$

Model	d_1	d_2	d_3	d_4	d_5
Empiric (3.2)	0.05	0.12	0.25	0.40	0.55
Logistic (3.3) with $a_0 = 0$	−2.94	−1.99	−1.10	−0.41	0.20
Logistic (3.3) with $a_0 = 3$	−5.94	−4.99	−4.10	−3.41	−2.80
Hyperbolic tangent (3.4)	−1.47	−1.00	−0.55	−0.20	0.10

The backward substitution (3.5) ensures the dose–toxicity model F
exact fit over the initial guesses of toxicity probabilities
the clinicians' prior beliefs. This is a crucial step partic
underparameterized (one-parameter) model. In practice, i
clinicians to provide reliable guesses for all test doses pr
take the approach by which the skeleton $\{p_{0k}\}$
good design's operating characteristics. We shall return to this in Chapter 8.

3.2.4 Model-Based MTD

The CRM starts a trial by treating the first patient at the prior MTD ν_0
$x_1 = d_{\nu_0}$. By (3.6), this starting dose is the dose initially believed
probability (closest to) θ. Each subsequent x_i
the previous observation history $H_i = \{(x_j, Y_j) : j < i\}$ for $i \geq 2$. A CRM design
can be viewed as a function defined on the increasing H_i
the next patient at the model-based MTD estimate, given H_i. Precisely,

$$x_i = \mathscr{D}_1(H_i) = \arg\min_{d_k} |F(d_k, \hat{\beta}_{i-1}) - \theta|$$

where

$$\hat{\beta}_{i-1} = \frac{\int_{-\infty}^{\infty} \beta L_{i-1}(\beta)\, dG(\beta)}{\int_{-\infty}^{\infty} L_{i-1}(\beta)\, dG(\beta)}$$

is the posterior mean of β given H_i and

$$L_{i-1}(\beta) = \prod_{j=1}^{i-1} \{F(x_j, \beta)\}^{Y_j} \{1 - F(x_j, \beta)\}^{1-Y_j}$$

is the binomial likelihood. The assignment (3.7) continues
until a prespecified sample size N
is given by $x_{N+1} = \mathscr{D}_1(H_{N+1})$, that is, a dose would have been given to the $(N+$
patient enrolled to the trial.

In other words, the CRM attempts to treat the next patient at t
of the MTD, a dose with toxicity probability estimated to be c
The motivation of this algorithm is to correct for the deficie
which treats the majority of the subjects at low and inefficac
is ethically sound on a conceptual level, there may be variou
"best" dose on the implementation level. The model-based MT
by estimating $p_k = \pi(d_k)$ with a plug-in estimate $F(d_k, \hat{\beta}_{i-1})$
estimate is

$$\mathscr{D}_1^*(H_i) = \arg\min_{d_k} |E_{i-1}\{F(d_k, \beta)\} - \theta|$$

where $E_{i-1}(\cdot)$ denotes expectation computed with respect to the posterior given
that is,

$$E_{i-1}\{F(d_k, \beta)\} = \frac{\int_{-\infty}^{\infty} F(d_k, \beta) L_{i-1}(\beta)\, dG(\beta)}{\int_{-\infty}^{\infty} L_{i-1}(\beta)\, dG(\beta)}.$$

The MTD estimate (3.9) involves the computation of K
is a more formal Bayesian estimate of p_k
the plug-in estimate (3.7) emerged to be the convention beca
ease (although computational consideration is of much less
the estimate (3.9) as a formal Bayesian estimate is conceptu
when F is a correct model for π
the model-based CRM is that its performance does not rely on c
F. Hence, we will focus on the plug-in CRM (3.7), which is studi
and systematically than the other estimators in the literature.

3.2.5 Normal Prior on β

Now, suppose that β has a normal prior distribution with mean $\hat{\beta}_0$
function (3.3), applying backward substitution (3.5) gives

$$d_k = \frac{\text{logit}(p_{0k}) - a_0}{\exp(\hat{\beta}_0)}$$

where $\text{logit}(p) = \log\{p/(1-p)\}$. As a result, we have

$$F(d_k, \beta) = \frac{\exp\left[a_0 + \exp(\beta - \hat{\beta}_0)\{\text{logit}(p_{0k}) - a_0\}\right]}{1 + \exp\left[a_0 + \exp(\beta - \hat{\beta}_0)\{\text{logit}(p_{0k}) - a_0\}\right]}.$$

Since $F(d_k, \beta)$ depends on the parameter β only via $\beta - \hat{\beta}_0$
normal, we may arbitrarily set $\hat{\beta}_0 =$
model (3.3) is therefore invariant to the mean of a normal pri
invariance property holds for the general class of dose–tox
Chapter 4.

3.2.6 Implementation in R

The R package 'dfcrm' consists of functions for the implemen
of the Bayesian CRM using the empiric (3.2) and logistic (3.3
the function `crm` takes cumulative patient data and returns a dose for the next
according to the model-based estimate (3.7).

```
> ### Return the recommended dose level for patient 6
> ### based on data from five patients
> library(dfcrm)
> p0 <- c(0.05,0.12,0.25,0.40,0.55)  # initial guesses
> theta <- 0.25                       # target toxicity rate
> y <- c(0, 0, 1, 0, 0)               # toxicity indicators
> lev <- c(3, 5, 5, 3, 4)             # dose levels
> fooB <- crm(p0,theta,y,lev,model="logistic",intcpt=3)
> fooB$estimate       # posterior mean of beta
[1] 0.2794614
> fooB$mtd
[1] 4
> fooB$doses          # Dose labels
[1] -5.944439 -4.992430 -4.098612 -3.405465 -2.799329
>
```

The above R code illustrates the usage of `crm` for a trial with $K =$
a target $\theta = 0.25$; the function is applied to observations from the first 5 su
receive dose levels 3, 5, 5, 3, and 4, where the third patient h
one-parameter logistic function (3.3) is used when the argument `model`

as "logistic". The default intercept value is 3, and can be mo
`intcpt`; thus, the specification "`intcpt=3`
When no value for the argument `model` is provided, `crm`
(3.2) as the default. The function computes the posterior mean of β
prior with mean 0. The prior standard deviation is specified by the argument
if `scale` is not specified (as in the last illustration), the default value is $\sqrt{1.34}$.

One-stage Bayesian CRM

The one-stage Bayesian CRM requires the specification of an array of *parameters* which can be classified into clinical parameters (or clinici parameters) and model parameters. Planning and implementa in a dose finding trial takes three steps:

1. *Setting the clinical parameters:*
   - Target toxicity probability θ
   - Number of test doses K
   - Prior MTD ν_0
   - Sample size N
2. *Calibrating the model parameters:*
   - Functional form of the dose–toxicity model $F(\cdot, \beta)$
   - Skeleton $\{p_{0k}\}$ and hence the dose labels d_ks via backward substitution
   - Prior distribution $G(\beta)$ of β
3. *Execution:* Treat the first patient at ν_0
   recent model-based MTD (3.7) as data accrue throughout the trial.

3.3 Two-Stage CRM

3.3.1 Initial Design

There are two practical difficulties associated with the use CRM. First, treating the first patient at the prior MTD rather level may raise safety concerns. Second, the use of a prior distribution $G(\beta)$ viewed as subjective and arbitrary.

To address the first difficulty, several authors suggest star lowest dose, and applying dose escalation restrictions whe appears aggressive. This approach can be represented by a tw as follows. First, specify an initial design as a predetermi sequence $\{x_{i,0}\}$ such that $x_{i-1,0} \leq x_{i,0}$. Then a two-stage CRM $\mathscr{D}_2(H_i)$

$$\mathscr{D}_2(H_i) = \begin{cases} x_{i,0}, & \text{if } Y_j = 0 \text{ for all } j < i, \\ \mathscr{D}_1(H_i) & \text{if } Y_j = 1 \text{ for some } j < i. \end{cases}$$

In other words, the initial design is in effect until the first

toxic outcome is observed, the trial turns to the model-base
ments. Because the 3+3 algorithm is familiar, there is an inc
"group-of-three" initial design by which escalation takes
three nontoxic outcomes. That is,

$$x_{1,0} = x_{2,0} = x_{3,0} = d_1, x_{4,0} = x_{5,0} = x_{6,0} = d_2, x_{7,0} = x_{8,0} = x_{9,0} = d_3 \ldots$$

and so on. There is, however, no clear justification for using
apart from convention. In fact, this initial design is somet
motivation of the CRM. We will study the calibration of the in
Chapter 10.

3.3.2 Maximum Likelihood CRM

In response to the difficulty associated with the subjectivi
we may use maximum likelihood estimation in conjunction wit
idea is simple and analogous to the Bayesian CRM: with data ob
$i-1$ patients, the dose for the next subject is computed as

$$x_i = \tilde{\mathscr{D}}_1(H_i) = \arg\min_{d_k} |F(d_k, \tilde{\beta}_{i-1}) - \theta|$$

where $\tilde{\beta}_{i-1} = \arg\max_\beta L_{i-1}(\beta)$ is the maximum likelihood estimate (mle) of
given H_i.

The R function crm
specification of the argument method:

```
> ### Compute the recommended dose level for patient 6
> ### using maximum likelihood CRM on the same data
> ### i.e., same values of p0, theta, y, lev
> fooL <- crm(p0,theta,y,lev,model="logistic",method="mle")
> fooL$estimate   # mle of beta
[1] 0.3142946
> fooL$mtd
[1] 5
>
```

The function crm evaluates the maximum likelihood estimate of MTD (3.11) thr
the specification of the method argument as "mle". When method
the Bayesian CRM is assumed.

Using (3.11) presupposes the existence of mle of β
F, the mle $\tilde{\beta}_{i-1}$ exists if and only if there is heterogeneity in the toxicity o
among patients, that is, $Y_j = 0$ and $Y_{j'} = 1$ for some $j, j' < i$
there is one toxic outcome out of 5 patients, and thus $\tilde{\beta}_5$
plans to use the maximum likelihood CRM, it is necessary to co
design, which can be formed by replacing $\mathscr{D}_1(H_i)$ with $\tilde{\mathscr{D}}_1(H_i)$ in (3.10).

The posterior mean $\hat{\beta}_{i-1}$ and the mle $\tilde{\beta}_{i-1}$

to potentially different dose recommendations. For exampl
the Bayesian CRM chooses dose level 4 for the sixth patient, w
likelihood CRM chooses level 5. However, the posterior mean $\hat{\beta}_5 = 0.$
$\tilde{\beta}_5 = 0.314$ are only slightly different. Looking at the model-based
toxicity probabilities reveals that both dose levels 4 and 5
from the target $\theta = 0.25$:

```
> round(fooB$ptox,digits=2)    # Posterior toxicity rates
[1] 0.01 0.03 0.08 0.18 0.33
> round(fooL$ptox,digits=2)    # Maximum likelihood estimates
[1] 0.01 0.02 0.07 0.16 0.30
>
```

The difference in the estimation of β
eventually, the choice of the estimation method per se will h
the dose assignments. However, the estimation method may ha
performance due to the fact that the maximum likelihood CRM a
two-stage strategy whereas Bayesian CRM is typically used a

Two-stage CRM

The two-stage CRM requires the specification of an initial de
design parameters in addition to those required in the one-s
and implementation of a two-stage CRM in a dose finding trial t

1. *Setting the clinical parameters:*
   - Target toxicity probability θ
   - Number of test doses K
   - Prior MTD ν_0
   - Sample size N
2. *Calibrating the model parameters:*
   - Functional form of the dose–toxicity model $F(\cdot, \beta)$
   - Skeleton $\{p_{0k}\}$ and hence the dose labels d_ks via backward substitution†
   - Prior distribution $G(\beta)$ of β, if Bayesian CRM is used
3. *Specifying an initial dose sequence:* $x_{1,0} \leq x_{2,0} \leq \ldots \leq x_{N,0}$.
4. *Execution:* Treat patients initially according to $\{x_{i,0}\}$
   toxic outcome, treat the subsequent patient at the most rece
   (3.7) or (3.11) as data accrue throughout the trial.

†Backward substitution for the maximum likelihood CRM can be
in (3.5) using an arbitrary initial value $\tilde{\beta}_0$
More details will be given in Chapter 4.

3.4 Simulating CRM Trials

3.4.1 *Numerical Illustrations*

Computer simulation is a primary tool for evaluating the agg
the method. To get a sense of how the CRM works, it is also usefu
individual simulated trials. Figure 3.1 shows the outcomes
a one-stage and two-stage CRM for a trial with target $\theta = 0.25$ and $K =$
true dose–toxicity scenario

$$p_1 = 0.02, p_2 = 0.04, p_3 = 0.10, p_4 = 0.25, p_5 = 0.50.$$

Hence, dose level 4 is the true MTD. The dose assignments by th
figure are quite different, although both select the correct
on the 20 simulated subjects. The one-stage CRM, taking adva
dose, treats the majority of the subjects at the MTD but also s
overdose. In contrast, the two-stage CRM takes a convention
initially and potentially treats many patients at low doses
CRM causes two toxic outcomes (patients 3 and 9) more than the
It is debatable whether the one-stage CRM is unsafe. On the on
been criticized on account of ethical concerns, as it is show
outcomes than the standard 3+3 method under some dose–toxic
On the other hand, overdosing is not necessarily a worse mist
when treating patients with severe diseases such as cancer.
a case study in which starting at a low dose (per the 3+3 algori
dilemma when consenting a cancer patient. Overall, the risk
be considered on a case-by-case basis. While the bortezomib
at the third level, the NeuSTART (Example 2.1) exercised cau
dose. In addition, the one-stage CRM in Figure 3.1 causes a 25
rate (5 out of 20), which is on target. From a numerical viewpo
design appears to be overconservative in this particular simulated trial.

3.4.2 *Methods of Simulation*

When simulating toxicity outcomes in a trial, each patient m
rying a latent toxicity tolerance u_i that is uniformly distributed on the interval [
If the uniform variate is smaller than the true toxicity prob
to the patient, the patient has a toxic outcome; otherwise, t
toxic outcome. That is,

$$Y_i = \begin{cases} 1 & \text{if } u_i \leq \pi(x_i) \\ 0 & \text{otherwise} \end{cases}$$

Table 3.2 displays the latent tolerance of the 20 simulated p
simulated trials in Figure 3.1. Consider, for example, pati
level 3 according to the one-stage CRM (left panel of the figur
toxic outcome because the tolerance $u_1 = .571 > p_3 = .$

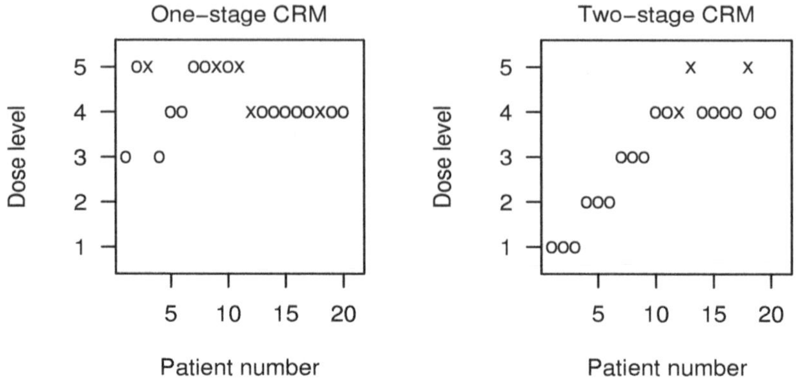

Figure 3.1 Simulated trials using the one-stage and two-sta
The logistic model (3.3) with $a_0 = 3$ in Table 3.1 is used with $\beta \sim N(0, 1.34)$
two-stage CRM, the initial design escalates according to th
represents a patient, with "o" indicating no toxicity and "x" indicating toxicity.

same CRM model as in Figure 3.1, we obtain the posterior mean $\hat{\beta}_1 = 0.$
implies $x_2 = 5$. Table 3.2 also gives the numerical outputs of the one-stag
based on these 20 simulated patients.

Table 3.2 A simulated CRM trial through latent toxicity tole

i	x_i	$\pi(x_i)$	u_i	y_i	$\hat{\beta}_i$	i	x_i	$\pi(x_i)$	u_i	y_i	$\hat{\beta}_i$
1	3	0.10	.571	0	0.60						
2	5	0.50	.642	0	0.93						
3	5	0.50	.466	1	0.04						
4	3	0.10	.870	0	0.18						
5	4	0.25	.634	0	0.28						
6	4	0.25	.390	0	0.34						
7	5	0.50	.524	0	0.41						
8	5	0.50	.773	0	0.47						
9	5	0.50	.175	1	0.31						
10	5	0.50	.627	0	0.35						

In a real trial, the latent tolerance u_i
on the other hand, toxicity tolerance can be easily generate
make different designs comparable in experiments where the
made adaptively. For example, the same latent tolerance seq
be used to generate the two-stage CRM in Figure 3.1 (right pan

one-stage and two-stage designs are treating the same patie
not necessarily treated at the same doses. The concept of tox
instrumental to the construction of a nonparametric optima
this in Chapter 5.

The function `crmsim`
CRM trials under a given dose–toxicity curve. The following
using the one-stage Bayesian CRM specified in Figure 3.1 unde
scenario (3.12):

```
> ### Generate 10 CRM trials
> theta <- 0.25
> PI <- c(0.02,0.04,0.10,0.25,0.50)  # True MTD = 4
> N <- 20                            # sample size
> x0 <- 3                            # starting dose level
> foo10 <- crmsim(PI,p0,theta,N,x0,nsim=10,model="logi
simulation number: 1
simulation number: 2
simulation number: 3
simulation number: 4
simulation number: 5
simulation number: 6
simulation number: 7
simulation number: 8
simulation number: 9
simulation number: 10
> foo10$MTD  # Display the distribution of recommended MTD
[1] 0.0 0.0 0.1 0.8 0.1
>
```

In this illustration, dose level 4 is selected as the MTD in 8 o

3.5 Practical Modifications

3.5.1 Dose Escalation Restrictions

The one-stage CRM in Figure 3.1 assigns dose level 5 to the sec
nontoxic outcome in the first patient who receives dose level
raise safety concerns because a high dose is tested without t
dose level. Several authors have noted the potential proble
the CRM, and proposed the restricted CRM by imposing an escal
The dose level for the next patient cannot be more than one lev
the current patient. Likewise, it is possible for the unrest
deescalation; cf. patient 4 under the one-stage CRM in Figur
doses in deescalation has not been perceived to be as problem
apply a similar restriction that the dose level for the next p
one level lower than that of the current patient. At any rate,
dose skipping will typically be applied, if ever, only to the

the change in the model-based estimates diminishes as the nu increases (try plotting the sequence of $\hat{\beta}_i$ in Table 3.2).

Another pathology in escalation is illustrated in the two-s the dose for patient 13 is escalated from that of patient 12, w Such an escalation is called incoherent as it puts patient 13 in light of the outcome in the previous patient [16]. To avoid moves by the CRM, we may apply the restrictions:

Coherence in escalation:
the dose level for the next patient cannot be higher than that patient.
Coherence in deescalation:
then the dose level for the next patient cannot be lower than t patient.

It is noteworthy that there is no incoherent move by the one-s That is, escalation occurs only after a nontoxic outcome, an toxic outcome. In fact, the one-stage CRM will never induce a even if no restriction is applied. This property indicates t is doing the right thing ethically. In contrast, an unrestri yield incoherent escalation as seen in Figure 3.1. It turns o to poor calibration of the initial design. In this particula escalation sequence is not an appropriate choice. This topi in Chapters 5 and 10.

Figure 3.2 shows the outcomes of the *restricted* specifications as in Figure 3.1 and the same 20 patients as in T assignment patterns are similar to that of the unrestricted the dose escalation restrictions have little impact on the C in terms of the probability of correctly selecting the MTD an number; try Exercise 3.4. Instead, these restrictions modify the of the design: a pointwise property of a dose finding design co of individual outcome sequences ("points"). The pointwise should agree with principles deemed sensible to clinicians principles (e.g., making incoherent moves) will be perceiv poor statistical properties. After all, only a single outco an actual trial. The aggregate operating characteristics a clinicians.

In the rest of this book, unless specified otherwise, we assum of no dose skipping in escalation and the coherence restrict is the assumption used in the function `crmsim` `restrict`.

3.5.2 Group Accrual

The fully sequential CRM follows the current patient over an called an *observation window*

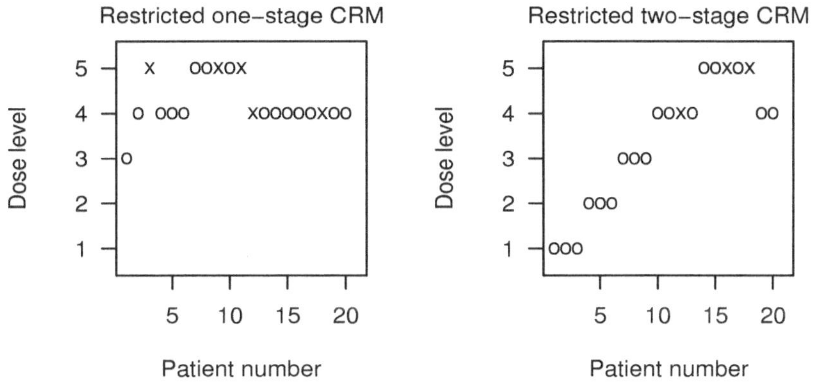

Figure 3.2 Simulated trials using the one-stage and the two-
escalation restrictions. Each point represents a patient,
indicating toxicity.

window (e.g., 6 months) will likely result in repeated inter
impose excessive administrative burdens, and cause long trial duration.

Goodman et al. [43] address this problem by assigning $m >$
to each dose, and updating the model-base estimate (3.7) bet
m patients. Figure 3.3 shows the outcomes of the simulated tri
CRM, using the models in Figure 3.1 and the same 20 patients in
left panel of Figure 3.3, the model-based MTD estimate (3.7)
$m = 2$ patients. The dose assignment pattern is similar to that of
CRM in Figure 3.1: half of the study subjects receive the MTD.
group size with $m =$
occasions to adapt to the previous outcomes. This may partly
the accuracy of MTD estimation decreases as a large group siz
assumed model F is not a correct specification of the true dose–toxicity curv
Therefore, although a large group size reduces the number of
and shortens trial duration, it also undermines the advanta
the CRM.

When the observation window is long in comparison to the recr
the reduction in trial duration by group accrual may not be ad
instance, each patient in the current group is to be followed
next group is enrolled in Figure 3.3. Assuming instantaneou
group CRM with size 4 will take roughly 24 months to enroll 20 p
group size of 2 will take 54 months. We will return to consider
a long observation window due to late toxicities in Chapter 1
the time-to-event continual reassessment method (TITE-CR
the group accrual CRM.

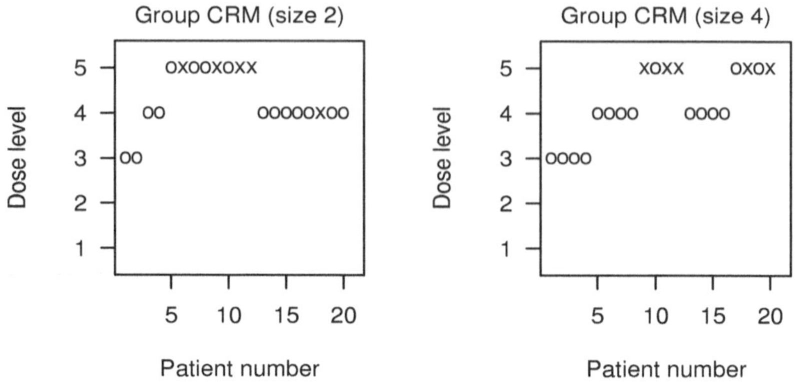

Figure 3.3 Simulated trials using the one-stage group accru
updated after every two patients in the left panel, and every
Dose escalation restrictions are applied. Each point repre
no toxicity and "x" indicating toxicity.

3.5.3 Stopping and Extension Criteria

It is common to have provisions for early termination in clin
and ethical reasons. The standard 3+3 design for dose finding
two toxic outcomes are observed at a dose. In contrast, the or
a fixed sample size N
reason is sufficiently compelling for stopping a CRM trial ea
sample size of a phase I trial is already small, and there is no
the sample size unless the early stopping rules are liberal.
expected to converge to the MTD as the trial accrues patients
enrolled towards the end of the trial will likely receive a go
will be a reduced ethical imperative to stop the trial.

On the other hand, there may be good reasons for extending enr
the CRM in some situations. Goodman et al. [43] consider cont
beyond N until the recommended MTD has at least a certain number of pat
assigned to it. This is to avoid the case where only few patien
recommended MTD. Consider the two-stage CRM in Figure 3.2 (r
on the 20 observations shown in the figure, the recommended MT
which is one level higher than the dose given to the final patie
patients with an observed toxicity rate of 40%. These are sig
MTD is not adequately assessed. Figure 3.4 shows the outcome
CRM trial with a minimum sample size criterion: the trial wil
until at least 9 patients have been treated at the recommende
of this modification, the trial continues for another 7 patie

subjects, there is evidence that dose level 5 exceeds the acc
(44% observed toxicity rate) and the final MTD is dose level 4.

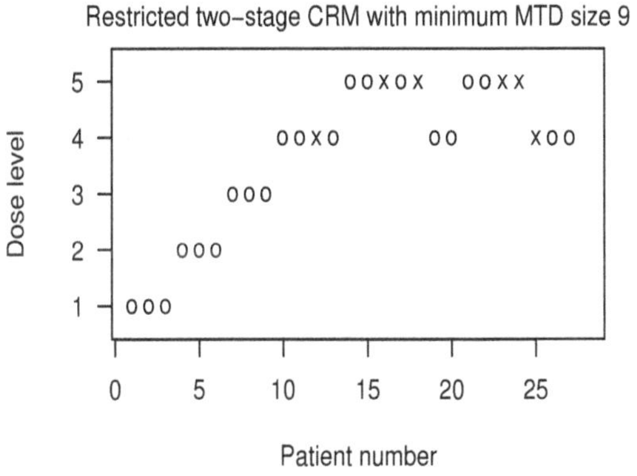

Figure 3.4 Simulated trials using the restricted two-stage
criterion. Dose escalation restrictions are applied. Each
indicating no toxicity and "x" indicating toxicity.

3.6 Bibliographic Notes

O'Quigley and Chevret [75] and Chevret [25] examine the oper
of the one-stage Bayesian CRM by simulation study. Moller [6
Shen [80] study the two-stage CRM; the latter also propose th
likelihood estimation. Several authors, including Faries
Goodman et al. [43], note the potential problem with dose ski
and propose to modify the CRM to limit escalation by no more th
time. Faries [36] also suggests enforcing coherence in esca
notion of coherence is subsequently formalized in Cheung [1
various variants of the CRM by simulations. Early stopping o
in Heyd and Carlin [45] and O'Quigley and Reiner [79]. A recen
CRM is given in Garrett-Mayer [39] and Iasonos et al. [48].

3.7 Exercises and Further Results

Exercise 3.1. Apply backward substitution (3.5) to the empiric function (
$\hat{\beta}_0 = 0$, and show that

$$F(d_k, \beta) = p_{0k}^{\exp(\beta)}$$

for a given skeleton $\{p_{01}, \ldots, p_{0K}\}$.

Exercise 3.2. Use the toxicity tolerance in Table 3.2 to verify the outcome
two-stage CRM in Figure 3.1.

Exercise 3.3. The function `crmsim`
fying the argument `x0`
specified sample size `N`
using the two-stage CRM in Figure 3.1 with a group-of-three initial rule.

Exercise 3.4. By default, the function `crmsim`
size 1. These options can be modified through the arguments `restrict`
`mcohort`. Usage of these options is documented in the reference manua
at the R Web site:

`http://www.r-project.org.`

Run simulations, and examine the effects of escalation rest
of recommended MTD under dose–toxicity curve (3.12).

Chapter 4

4.1 Introduction

One-parameter functions are predominantly used in the CRM l
dose–toxicity relationship. An introduction on the use of o
in order—for two reasons. First, these functions are underp
be expected to produce a realistic fit on the entire dose range
be flexible enough to provide a reasonable approximation *locally*
dose. Second, the use of one-parameter models is not common i
areas, and modeling in the context of CRM is quite different f
approach. This chapter presents a unified framework for CRM m
introduces a class of dose–toxicity functions that include
CRM models. The required assumptions on a one-parameter CRM
in Section 4.3. Proof of the main result in this chapter is given in Section 4.4.

4.2 ψ-Equivalent Models

Consider the following class of dose–toxicity functions:

$$F(x,\beta) = \psi\{c(\beta)h(x)\}$$

where the parameter β is a scalar, and the functions ψ, c, and h
and known. This class of functions includes the most commonl
curves in the CRM literature. For example, the empiric funct

$$\psi(z) = \exp(z),\, c(\beta) = \exp(\beta),\, h(x) = \log(x)$$

so that

$$F(x,\beta) = \exp\{\exp(\beta)\log(x)\} = x^{\exp(\beta)}.$$

Also, the logistic function (3.3) with a fixed intercept a_0 is represented by

$$\psi(z) = \frac{\exp(a_0+z)}{1+\exp(a_0+z)},\, c(\beta) = \exp(\beta),\, h(x) = x.$$

Recall that the dose labels in the CRM are obtained by backwar
that $F(d_k,\hat{\beta}_0) = p_{0k}$, where p_{0k}

level k, and that $\hat{\beta}_0$ is the prior mean of β
in a maximum likelihood CRM. Therefore, a CRM model is defined
form $F(\cdot,\beta)$ of the dose–toxicity curve, the skeleton $\{p_{01},\ldots,p_{0K}\}$
value $\hat{\beta}_0$. (However, see Theorem 4.1 below.) Denote

$$F_k(\beta) = F(d_k,\beta) \text{ for } k = 1,\ldots,K.$$

Under (4.1), the *ψ-representation* of a CRM model can be written as

$$F_k(\beta) = \psi\{c(\beta)h(d_k)\} = \psi\left\{\frac{c(\beta)}{c(\hat{\beta}_0)}\psi^{-1}(p_{0k})\right\}$$

which depends on the dose index k via the initial guess p_{0k}
distinguish between doses, therefore, the skeleton needs to be specified as a
increasing sequence. Furthermore, the representation (4.
function $h(x)$. Thus, two CRM models are identical if their dose–toxicity f
are represented by the same $\psi(z)$ and $c(\beta)$
function (3.4) corresponds to

$$\psi(z) = \exp(z) \text{ and } c(\beta) = \exp(\beta)$$

with $h(x) = \log\{(\tanh x + 1)/2\}$. Since the function is represented by the same
and $c(\beta)$ as in (4.2), the CRM model defined by the hyperbolic tangent fu
identical to that by the empiric function. The ψ
is

$$F_k(\beta) = \exp\left\{\frac{\exp(\beta)}{\exp(\hat{\beta}_0)}\log(p_{0k})\right\} = p_{0k}^{\exp(\beta-\hat{\beta}_0)}.$$

Example 4.1. The logistic function with a fixed slope $a_1 > 0$:

$$F(x,\beta) = \frac{\exp(\beta + a_1 x)}{1+\exp(\beta + a_1 x)}$$

corresponds to $\psi(z) = z/(1+z)$ and $c(\beta) = \exp(\beta)$
of a_1. After backward substitution, the model can be represented by

$$F_k(\beta) = \frac{\exp(\beta - \hat{\beta}_0)p_{0k}}{1 - p_{0k} + \exp(\beta - \hat{\beta}_0)p_{0k}}.$$

The function (4.4) leads to the same CRM model regardless of the value of a_1
may therefore set $a_1 = 1$ arbitrarily.

Example 4.1 illustrates how the ψ
CRM models, and reduce unnecessary comparisons of different functions (e.g.,

Definition 4.1 (ψ-equivalent models).
to be ψ-equivalent if their dose–toxicity functions can be repres
$\psi(z)$.

ψ-EQUIVALENT MODELS

While ψ-equivalent models are not necessarily identical, the diff
a result of different parameterizations. For example, anot
function,

$$F(x,\beta) = x^{\beta},$$

can be represented by $\psi(z) = \exp(z)$ and $c(\beta) = \beta$, and is ψ
While (3.1) and (3.2) are not equivalent, two functions diff
the parameter appears in the functions. Hence, both functio
estimation of the toxicity probabilities when maximum like

Theorem 4.1. *Suppose that $F_k^{(1)}(\beta)$ and $F_k^{(2)}(\phi)$ are derived from two ψ*
models and the same skeleton $\{p_{0k}\}$. Let $\tilde{\beta}_{i-1}$ and $\tilde{\phi}_{i-1}$
likelihood estimates of β and ϕ
Then

$$F_k^{(1)}(\tilde{\beta}_{i-1}) = F_k^{(2)}(\tilde{\phi}_{i-1})$$

for all k.

In words, Theorem 4.1 states that the maximum likelihood CRM
a ψ-equivalent class of CRM models. The proof is an extension of
property of maximum likelihood estimation, and is given in S
Theorem 4.1 holds regardless of the initial values of β and ϕ
can arbitrarily choose the initial values, $\hat{\beta}_0$ and $\hat{\phi}_0$
step (3.5). Without loss of generality, from now on, we will s
that $c(\hat{\beta}_0) = c(\hat{\phi}_0) = 1$.

Example 4.2. The dose–toxicity function

$$F(x,\phi) = \frac{\phi x^2}{a_2 + \phi x^2} \quad \text{for some } a_2 >$$

can be represented by $\psi(z) = z/(1+z)$ with $c(\phi) = \phi$ and $h(x) = x^2/a_2$
the CRM model generated by (4.6) is ψ
with a fixed slope; cf., equation (4.4) in Example 4.1. That is
identical dose assignments if maximum likelihood CRM is used.

In contrast, the Bayesian CRM (3.7) is not invariant among ψ
The model (4.6) in Example 4.2 can be represented by

$$F_k(\phi) = \frac{(\phi/\hat{\phi}_0)\, p_{0k}}{1 - p_{0k} + (\phi/\hat{\phi}_0)\, p_{0k}}$$

where $\hat{\phi}_0$ is the prior mean of ϕ
lead to identical posterior computations if $\exp(\beta - \hat{\beta}_0)$ and $\phi/\hat{\phi}_0$
distribution. However, this is impossible because $E_0(\phi/\hat{\phi}_0) =$
whereas

$$E_0\left\{\exp(\beta - \hat{\beta}_0)\right\} > \exp\left\{E_0(\beta - \hat{\beta}_0)\right\} = 1$$

by Jensen's inequality. This example illustrates that parameterization via $c(\beta)$ an important consideration when the Bayesian CRM (3.7) is used.

When $c(\beta) = \exp(\beta)$, the ψ
the parameter β only via $\beta - \hat{\beta}_0$. If the prior distribution of β
variable $\beta - \hat{\beta}_0$ is always mean zero normal. Therefore, the posterior comput
will be identical regardless of the specified prior mean $\hat{\beta}_0$
invariant to the mean of a normal prior. In general, this inva
Bayesian CRM when model (4.1) is used and the prior distribution of β
location-scale family (Lehmann, 1983, page 20). In this boo
model class (4.1) with $c(\beta) = \exp(\beta)$ and a normal prior distribution for β
model parameters of the Bayesian CRM are the function $\psi(z)$, the skeleton $\{p$
and the variance σ_β^2
$\psi(z)$ and the corresponding CRM models. The calibration of $\{p_{0k}\}$
in Chapter 8, and specification of σ_β^2 in Chapter 9.

4.3 Model Assumptions

In this section, we state the regularity conditions on the on
$F_k(\beta)$ assumed in this book. The role of these conditions in various
will be discussed in the following chapters. In this chapter
behind the conditions and their implications. A practical p
are verifiable for any given $F_k(\beta)$
remove certain models from consideration for use.

Condition 4.1. *$F(x, \beta)$ is strictly increasing in x for all β.*

Condition 4.2. *$F_k(\beta)$ is monotone in β in the same direction for all k.*

Conditions 4.1 and 4.2 are satisfied by many dose–toxicity fu
the model class (4.1) satisfies these conditions when either $h(d_k) > 0$ for all k or
for all k. It is easy to verify that the logistic model with $a_0 =$
satisfy Condition 4.2 because $h(d_k) < 0$ for $k \leq 4$ but $h(d_5) >$
should not be considered for use. Graphically, Condition 4.
of curves induced by $F(x, \beta)$ do not cross each other; see Figure 4.1.

For the next condition, we first define $g_{ij}(\beta) = \{1 - F_i(\beta)\}/\{1 - F_j(\beta)\}$.

Condition 4.3. *The derivatives $F_k'(\beta)$ and $g_{ij}'(\beta)$ exist and $g_{ij}'(\beta)F_k'(\beta) \leq 0$
k and $i > j$.*

Condition 4.3 can be equivalently stated as

$$|F_i'(\beta)| \geq |F_j'(\beta)| \{1 - F_i(\beta)\}| / \{1 - F_j(\beta)\}$$

for all $i > j$, which puts a lower bound on $|F_i'(\beta)|$ relative to $|F_j'(\beta)|$
that $\log\{1 - F_k(\beta)\} = \int_{-\infty}^{\beta} [-F_k'(\phi)/\{1 - F_k(\phi)\}] d\phi$ when $F_k'(\phi) > 0$ for all k
can show that Condition 4.3 implies Condition 4.1 and hence i

Table 4.1 Dose–toxicity functions and their ψ–representations with $c(\beta) = \exp(\beta)$, $h(x) = x$, and $\hat{\beta}_0 = 0$

Model	$\psi(z)$	$F_k(\beta)$	Range of β
Empiric (or hyperbolic tangent)	$\exp(z)$	$p_{0k}^{\exp(\beta)}$	$(-\infty, \infty)$
Complementary log-log	$1 - \exp(-z)$	$1 - (1 - p_{0k})^{\exp(\beta)}$	$(-\infty, \infty)$
Logistic; fixed intercept a_0	$\frac{\exp(a_0 + z)}{1 + \exp(a_0 + z)}$	$\frac{\exp[a_0 + \exp(\beta)\{\text{logit}(p_{0k}) - a_0\}]}{1 + \exp[a_0 + \exp(\beta)\{\text{logit}(p_{0k}) - a_0\}]}$	$(-\infty, \infty)$
Logistic; fixed slope	$\frac{z}{1 + z}$	$\frac{\exp(\beta) p_{0k}}{1 - p_{0k} + \exp(\beta) p_{0k}}$	$(-\infty, \infty)$
Product-of-beta (12.10)	$1 - \frac{1}{z}$	$1 - \exp(-\beta)(1 - p_{0k})$	$> \log(1 - p_{01})$
Probit; fixed intercept a_0	$\Phi(a_0 + z)$	$\Phi\left[a_0 + \exp(\beta)\left\{\Phi^{-1}(p_{0k}) - a_0\right\}\right]$	$(-\infty, \infty)$
Probit; fixed slope	$\Phi\{\log(z)\}$	$\Phi\{\beta + \Phi^{-1}(p_{0k})\}$	$(-\infty, \infty)$

Note: $logit(p) = \log\{p/(1-p)\}$; $\Phi(z)$ *indicates cdf of standard normal at z.*

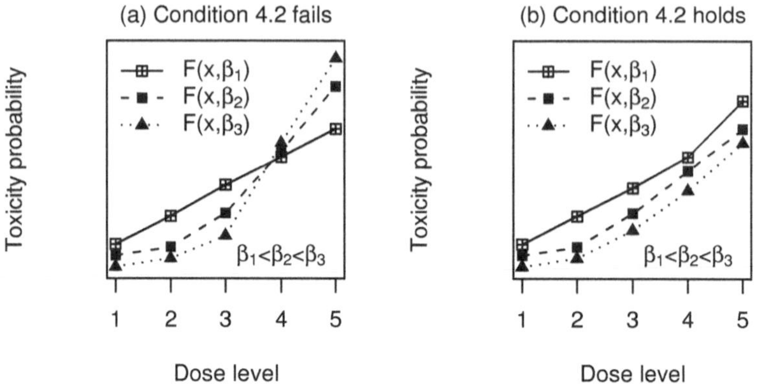

Figure 4.1 Graphical illustration of Condition 4.2. (a) A model $F_k(\beta)$ for $k = 1,2,3$, and increasing in β for $k = 4$, $F_k(\beta)$ that is uniformly decreasing in β for all dose level k.

Example 4.3.
Conditions 4.1 and 4.2, because (4.5) is increasing in k and increasing in β all k. To further show that the model satisfies Condition 4.3, we fir $F_k'(\beta) = F_k(\beta)\{1-F_k(\beta)\} > 0$. Next, for $i > j$,

$$
\begin{aligned}
g_{ij}'(\beta) &= \frac{F_j'(\beta)\{1-F_i(\beta)\} - F_i'(\beta)\{1-F_j(\beta)\}}{\{1-F_j(\beta)\}^2} \\
&= \frac{F_j(\beta)\{1-F_j(\beta)\}\{1-F_i(\beta)\} - F_i(\beta)\{1-F_i(\beta)\}\{1-F_j(\beta)\}}{\{1-F_j(\beta)\}^2} \\
&= \frac{\{F_j(\beta)-F_i(\beta)\}\{1-F_i(\beta)\}\{1-F_j(\beta)\}}{\{1-F_j(\beta)\}^2} < 0.
\end{aligned}
$$

Hence, $g_{ij}'(\beta)F_k'(\beta) < 0$ and Condition 4.3 is met.

Condition 4.4. *For any given $\theta \in (0,1)$ and a given ν, there exists β eter space such that $F_\nu(\beta) = \theta$.*

Condition 4.5. *Suppose $F_k(\beta)$ is decreasing in β. There exist the constants $\underline{b}$ such that $F_1(\underline{b}) > \theta$ and $F_K(\bar{b}) < \theta$ and $[\underline{b}, \bar{b}]$ space of β.*

Consider the ψ-representation of the empiric model in Table 4.1. Conditio satisfied because we can set $\beta = \log\{\log(\theta)/\log(p_{0\nu})\} \in (-\infty, \infty)$ and $\theta \in (0,1)$. Also, Condition 4.5 is met by setting

$$\underline{b} < \log\{\log(\theta)/\log(p_{01})\} \text{ and } \bar{b} > \log\{\log(\theta)/\log(p_{0K})\}.$$

Intuitively, Conditions 4.4 and 4.5 ensure that the model is
approximate dose–toxicity scenarios with all possible MTD ν and any given
that there always exists x^* that solves the equation $F(x^*, \beta) = \theta$
x^* is guaranteed by Condition 4.1.

Condition 4.5 assumes $F_k(\beta)$ is decreasing in β
$F_k(\beta)$ that is increasing in β can be also stated.

Condition 4.6. *$F_k(\beta)$ is bounded away from 0 and 1 on $[\underline{b}, \overline{b}]$ for all k; and F is uniformly bounded in β.*

Condition 4.7. *For any given $0 < p < 1$ and each k, the function*

$$p\frac{F_k'(\beta)}{F_k(\beta)} + (1-p)\frac{-F_k'(\beta)}{1-F_k(\beta)}$$

is continuous and strictly monotone in β.

Condition 4.8. *For $\beta \in [\underline{b}, \overline{b}]$,*

$$F_k(\beta)F_k''(\beta) - \{F_k'(\beta)\}^2 \leq$$

and

$$-F_k''(\beta) + \left[F_k(\beta)F_k''(\beta) - \{F_k'(\beta)\}^2\right] \leq$$

with at least one inequality being strict for all k.

Condition 4.6 is introduced for technical brevity (in the pr
can be relaxed by modifying the tails of the function F
is met by any dose–toxicity functions generated by a continuous $\psi(z)$
class (4.1). Condition 4.8 is an additional assumption requ
of the Bayesian CRM using posterior mean, and is not required
likelihood CRM or Bayesian CRM using *posterior mode* (cf. Exercise 4.4).

If a dose–toxicity function satisfies the above conditions, its ψ
will also satisfy the conditions. Therefore, verification o
done under a parameterization that allows simple algebraic manipulations.

Example 4.4. To verify that the empiric model (3.2) satisfies Condition 4.
may instead work with model (3.1) for which $F_k'(\beta) = F_k(\beta)\log d_k$ and $F_k''($
$F_k(\beta)(\log d_k)^2$. Hence, $F_k(\beta)F_k''(\beta) - \{F_k'(\beta)\}^2 = 0$,
$-F_k''(\beta) < 0$, which satisfies (4.9) and the inequality is strict under Co

One may also prove the empiric model satisfies Condition 4.8 u
form (3.2) directly; see Exercise 4.3.

4.4 Proof of Theorem 4.1

To prove Theorem 4.1, we first express the two models as

$$F_k^{(1)}(\beta) = \psi\left\{\frac{c_1(\beta)}{c_1(\tilde{\beta}_0)}\psi^{-1}(p_{0k})\right\} \text{ and } F_k^{(2)}(\phi) = \psi\left\{\frac{c_2(\phi)}{c_2(\tilde{\phi}_0)}\psi^{-1}(p_{0k})\right\},$$

where $\tilde{\beta}_0$ and $\tilde{\phi}_0$ are the respective initial values used in backward substitu two models. Now, since the models depend on the parameters through $c_1(\beta)/c_1($ and $c_2(\phi)/c_2(\tilde{\phi}_0)$

$$\frac{\widetilde{c_1(\beta)}}{c_1(\tilde{\beta}_0)} = \frac{\widetilde{c_2(\phi)}}{c_2(\tilde{\phi}_0)}, \text{ which implies } \frac{c_1(\tilde{\beta}_{i-1})}{c_1(\tilde{\beta}_0)} = \frac{c_2(\tilde{\phi}_{i-1})}{c_2(\tilde{\phi}_0)}$$

for a given H_i by invariance of maximum likelihood estimation. It thus fol $F_k^{(1)}(\tilde{\beta}_{i-1}) = F_k^{(2)}(\tilde{\phi}_{i-1})$.

4.5 Exercises and Further Results

Exercise 4.1. Verify Conditions 4.1–4.3 for the empiric model (3.2).

Exercise 4.2. Verify Conditions 4.4 and 4.7 for the logistic model (3.3) wi intercept a_0. For given θ and p_{0k}s, determine the range of a_0 is met.

Exercise 4.3. Verify Condition 4.8 for the empiric model using the form (3. compare your result with Example 4.4.

Exercise 4.4 (invariance of CRM using posterior mode). assigns the doses based on the posterior mode of β, that is,

$$x_i = \arg\min_{d_k} |F_k(\check{\beta}_{i-1}) - \theta|,$$

where $\check{\beta}_{i-1}$ is the maximum of the posterior density given H_i. Suppose that $F_k^{(1)}$ and $F_k^{(2)}(\phi)$ are ψ-equivalent models with respect to the mode, and can be writt

$$F_k^{(1)}(\beta) = \psi\left\{\frac{c_1(\beta)}{c_1(\check{\beta}_0)}\psi^{-1}(p_{0k})\right\} \text{ and } F_k^{(2)}(\phi) = \psi\left\{\frac{c_2(\phi)}{c_2(\check{\phi}_0)}\psi^{-1}(p_{0k})\right\}.$$

Let $G_1(\beta)$ and $G_2(\phi)$ denote the prior distributions of β and ϕ. Prove that

$$F_k^{(1)}(\check{\beta}_{i-1}) = F_k^{(2)}(\check{\phi}_{i-1}) \text{ for all } k$$

if

$$G_1\left[c_1^{-1}\left\{c_1(\check{\beta}_0)z\right\}\right] = G_2\left[c_2^{-1}\left\{c_2(\check{\phi}_0)z\right\}\right] \text{ for all } z.$$

A direct consequence of (4.10) is that the CRM using the two models $F^{(1)}$ and will be identical if the prior distributions are matched. Si established for CRM using the median.

Chapter 5

Theoretical Properties

5.1 Introduction

This chapter presents some theoretical criteria for the eva
methods, and discusses the properties of the CRM in light of t
the concept of coherence is introduced in Section 5.2. The as
the CRM is then discussed in Section 5.3, where we also demons
in finite sample settings. Finally, Section 5.4 sketches the
in this chapter.

5.2 Coherence

5.2.1 Motivation and Definitions

Although the 3+3 dose escalation algorithm has fundamental
MTD estimation, it reflects an important ethical principle w
for the next patient is appropriate only when the most recent
patients) does not show any sign of toxicity. This design fea
limits the risk of unduly exposing patients to high and toxic
most recent patient does not have a toxic outcome, it is count
the dose for the next patient, while the trial objective is to
to patients within safety limit. The use of outcome-adaptiv
studies indeed underlines these coherence considerations
be formulated as follows. Recall that Y_i
a fully sequential trial. An escalation for patient i
$Y_{i-1} = 0$; and a dose finding design $\mathscr{D}$
probability one,

$$P_{\mathscr{D}}(U_i > 0 | Y_{i-1} = 1) = 0 \text{ for all } i$$

where $U_i = x_i - x_{i-1}$ is the dose increment from patient $i-1$ to patient i, and
denotes the probability computed under the design $\mathscr{D}$
patient i is said to be coherent only when $Y_{i-1} = 1$; and a design $\mathscr{D}$
coherent in deescalation if with probability one

$$P_{\mathscr{D}}(U_i < 0 | Y_{i-1} = 0) = 0 \text{ for all } i$$

In other words, a coherent design does not induce any incoher
consider the unrestricted two-stage CRM in Figure 3.1. The d

for patient 13 to dose level 5 after a toxic outcome is observe
incoherent escalation, and by definition, this two-stage CR
On the one hand, one could argue that the incoherent escalati
be justified ethically because it exposes the patient to a hig
demonstrating safety at the lower doses. On the other hand, o
based on the outcomes in the first 11 patients, patient 12 woul
5 regardless of the outcome of patient 12; thus, with the obje
patients at the right dose as possible, it seems that the esca
for patient 12. Either way, the fact that a design is incohere
entirely in line with the ethics and objective of the trial.

5.2.2 Coherence Conditions of the CRM

The one-stage Bayesian CRM $\mathscr{D}_1$ is coherent.

This fact indicates the model-based CRM is intrinsically co
ethical principles in dose finding studies, although there i
coherence. The proof of the method's coherence is provided in Section 5.4.

In contrast, the two-stage CRM $\mathscr{D}_2$
two-stage design in Figure 3.1. Consider another two-stage
CRM model as in Figure 3.1, but with an initial design that esc
of two nontoxic outcomes. Figure 5.1 shows a simulated trial
by this design. There is no indication of incoherent moves. H
this design is indeed coherent, we need to verify that the des
incoherent move in all 2^{20} possible outcome sequences.

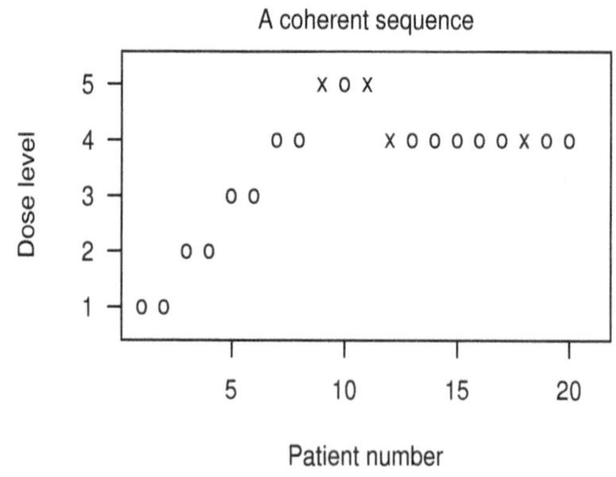

Figure 5.1 Simulated trials generated by a coherent two-sta
escalation restrictions. Each point represents a patient,
indicating toxicity.

Theorem 5.1 (coherence condition). *Let* $M_0 = \min\{i : Y_i = 1\}$ *first patient who experiences a toxicity. If* $\mathscr{D}_2(H_{M_0+1}) \leq \mathscr{D}_2(H_{M_0})$ *one, then the two-stage CRM* $\mathscr{D}_2$ *is coherent.*

In words, Theorem 5.1 states that a two-stage CRM is coherent induce an incoherent escalation at the transition from the i the model-based CRM (stage 2). Since there are only $N -$ transition may occur, this theorem reduces the computation checking significantly. For example, we can consider the outcome sequence

$$\mathbf{Y^{(i)}} = (Y_1^{(i)}, \ldots, Y_N^{(i)})$$

where $Y_j^{(i)} = 0$ for $j \neq i$, and $Y_i^{(i)} =$

$$\mathbf{X^{(i)}} = (X_1^{(i)}, \ldots, X_N^{(i)})$$

by a two-stage CRM $\mathscr{D}_2$. Then, the design $\mathscr{D}_2$ is coherent if

$$X_{i+1}^{(i)} \leq X_i^{(i)} \text{ for } i = 1, \ldots, N-1,$$

that is, no incoherent escalation is induced for patient $i+1$.

The R function `cohere`

checks the coherence status of a two-stage CRM. The design us is confirmed to be coherent using this function:

```
> ### Verify coherence of two-stage CRM in Figure 5.1
>
> p0 <- c(0.05,0.12,0.25,0.40,0.55)       # initial guesses
> theta <- 0.25                           # target DLT rate
> x0 <- c(1,1,2,2,3,3,4,4,rep(5,12))      # initial design
> foo <- cohere(p0,theta,x0,model="logistic")
> foo$message
[1] "Coherent"
>
```

Faries [36] notes the potential problem of incoherent escal which starts at a dose below the prior MTD, and proposes enfor *restrictions*. A natural question: what advantage checking the coherence in Theorem 5.1 offers over the simple use of dose escalation r answering this question, we will explore a related concept c

5.2.3 *Compatibility*

The initial design sequence $\{x_{i,0}\}$ when there is no observed toxicity throughout the trial. Int outcome occur, dose escalation may proceed at a rate slower t this initial sequence. With this in mind, the sequence $\{x_{i,0}\}$ reflect the "fastest" escalation pace for a target toxicity rate θ of escalation will be made specific in Chapter 10.)

Definition 5.1 (compatibility). *An initial design* $\{x_{i,0}\}$ *the CRM in a two-stage CRM* $\mathscr{D}_2$ *in (3.10) with respect to* θ *if* $\mathscr{D}_2(H_i) \leq x_{i,0}$ *with probability one.*

Figure 5.2 shows an outcome sequence of 20 patients by a two-s a group-of-four initial design. In the left panel, an escala after patient 14 has a toxic outcome. Furthermore, the toxic seems to have speeded up the pace of escalation: patient 15 wo level 4 if patient 11 did not experience toxicity. To mend the one may simply adopt dose escalation restrictions as discus right panel of Figure 5.2 indicates no violation of coherenc of incompatability remains: patient 16 would have received occurred prior to his.

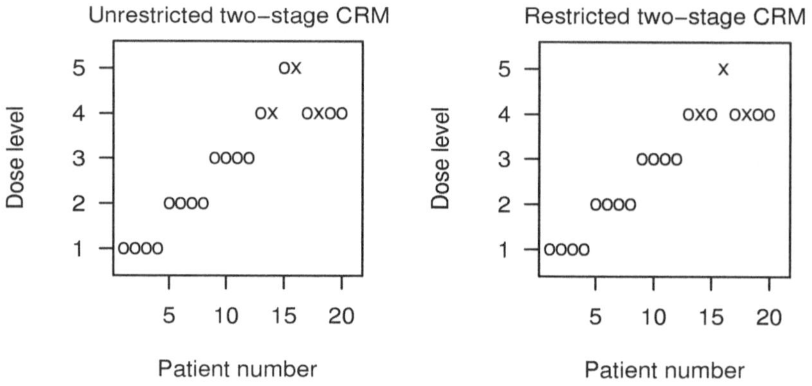

Figure 5.2 Simulated trials by an incompatible two-stage CR are applied in the right panel to enforce coherence, but not i patient, with "o" indicating no toxicity and "x" indicating toxicity.

The property of compatibility is more subtle than coherence against compatibility does not cause any immediate ethical cast doubt on the soundness of the overall design when the mod the initial sequence give conflicting dose assignments. The is always coherent whereas a two-stage CRM is not indicates t arise as a result of the improper choice of the initial design.

Theorem 5.2 (compatibility condition). *the initial design* $\{x_{i,0}\}$ *is chosen such that an unrestricted two-stage CRM* $\mathscr{D}$ *coherent, then* $\{x_{i,0}\}$ *is compatible with the CRM with respect to* θ*.*

Theorem 5.2 establishes compatibility as a consequence of c turn may be assessed by Theorem 5.1. In contrast, enforcing c

does not guarantee compatibility; cf., Figure 5.2. It motiv
our goal to calibrate the initial design so that the two-stag
applying any dose escalation restrictions.

A pragmatic approach may enforce coherence and compatibili
This is justified, in the sense that any statistical and model
be overruled if it contradicts sound clinical judgment. How
hoc restrictions on the CRM out of convenience, we may underm
a model-based design may offer. While model-based decision
sound clinical judgments, a model-based design should be ca
best mimic clinical judgments in a systematic manner, witho
ad hoc rules. In this regard, the criteria of coherence and co
tools to evaluate whether an initial design is appropriate f
used with a particular CRM model. As a case in point, in practi
use a group-of-three initial design with a two-stage CRM. Th
for this choice apart from its resemblance to the traditiona
the basis of coherence, we observe in Figure 3.1 that a group-
is not appropriate for $\theta = 0$.
pendence of coherence, the initial design, and the target θ
two-stage CRM design.

5.2.4 *Extensions*

For the general enrollment situations with $m \geq$
perimental unit and modify the coherence definitions in (5.1
as

$$P_{\mathscr{D}}(U_j > 0 | \bar{Y}_{j-1} \geq \theta) = 0 \text{ and } P_{\mathscr{D}}(U_j < 0 | \bar{Y}_{j-1} \leq \theta) = 0$$

for all j, where U_j in this context denotes the dose increment from group $j -$
j, and $\bar{Y}_j$ is the observed proportion of toxicity in group j
is group-coherent in escalation if it escalates only when th
in the most current cohort is strictly less than the target θ
group-coherent in deescalation if it deescalates only when
in the most current group is strictly greater than θ
size is equal to 1, the group-coherence conditions reduce to
[16] proves the group coherence of a one-stage CRM that assig
posterior mode of β with a unimodal prior.

The concept of group coherence makes practical sense only if
available between when the first patient in a group arrives an
of the group is treated. This assumption does not reflect the c
the toxicity outcome may occur at a random time during the obs
this case, a real-time extension of the coherence concept is
of coherence becomes complicated, however, it also becomes
See Chapter 11 for a discussion on the real-time coherence of
the CRM.

In a combined phase I/II trial, dose escalation decisions de
outcomes such as tumor shrinkage as well as toxicity outcome

are various ways to extend the CRM to use both the toxicity and
for dose finding, and a consensus is yet to be reached. However
coherence is quite straightforward. Namely, a deescalatio
current patient does not have a positive efficacy outcome. Se
explore the notion of bivariate coherence and its application.

5.3 Large-Sample Properties

5.3.1 Consistency and Indifference Interval

A primary estimation concern in a dose finding trial is whethe
correctly identified. If a design $\mathscr{D}$
select the correct dose when the sample size becomes sufficie
we are concerned about strong consistency of the design; that is, whether $\mathscr{D}(H_N$
d_ν eventually as $N \to \infty$ with probability one, where

$$\nu = \arg\min_k |p_k - \theta|.$$

Since $\mathscr{D}(H_N)$ denotes the dose assignment of patient N
connotation that as a trial keeps enrolling, all patients wi
the MTD.

However, it can be difficult for a design to achieve consisten
dose–toxicity scenarios, when there are other practical co
and discrete dosing. Alternatively, therefore, we may try t
eventual selected dose is to the target θ
such "closeness" is by the *indifference interval* of a dose finding design.

Definition 5.2 (indifference interval). *The interval* (θ_L, θ_U), *for* $\theta_L < \theta < \theta_U$
said to be an indifference interval of a dose finding design $\mathscr{D}$ *if for some* $N > 0$,

$$P_{\mathscr{D}}\{x_n \in \mathscr{I}_\pi(\theta_L, \theta_U) \text{ for all } n \geq N\} = 1,$$

where $\mathscr{I}_\pi(p_L, p_U) = \{x : p_L \leq \pi(x) \leq p_U\}$.

In other words, an indifference interval is an interval in wh
of the selected dose will eventually fall. In other words, a d
interval (θ_L, θ_U) may fail to differentiate the true MTD from a dose nearby, as l
as the toxicity probability of the dose falls within (θ_L, θ_U).

Not every dose finding design has an indifference interval; C
examples that do not have indifference intervals. A design is said to be δ
if its indifference interval can be expressed as $\theta \pm \delta$ for some $\delta \in (0, \theta)$
$\theta_L = \theta - \delta$ and $\theta_U = \theta + \delta$. If a design has an indifference interval (θ_L, θ_U)
not symmetric about θ, it is δ-sensitive by defining $\delta = \max(\theta_U - \theta, \theta - \theta_L)$.

The CRM is δ-sensitive. Furthermore, it can be calibrated to achieve any δ
of sensitivity. The smaller the half-width δ
design will have asymptotically: In Figure 5.3, the design with $\delta = 0.$
consistent estimates for the true MTD under both curves (bot

design with $\delta = 0$.
This having been said, a half-width 0.05 is arguably sufficiently small for $\theta =$
in many practical situations. Again, under Curve 2 in Figure 5.3, the design with
0.05 may not select the MTD, but will eventually select a dose in
In addition, since indifference interval is an asymptotic concept, a smaller δ
necessarily a better choice than a larger value in finite samp
to the choice of δ in Chapter 8.

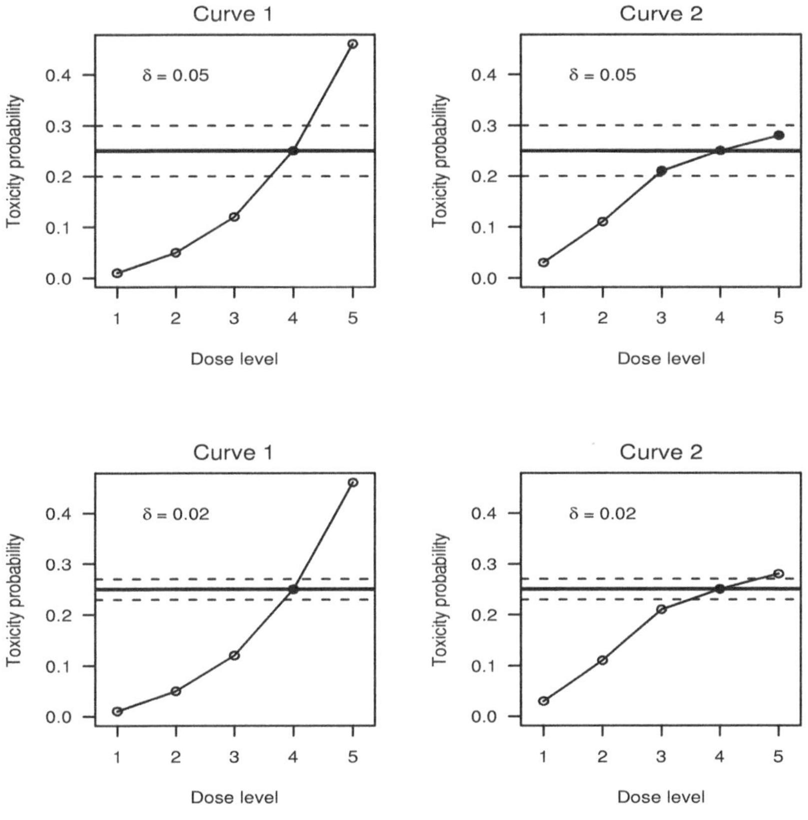

Figure 5.3 Two dose–toxicity curves under which dose level 4 is the true MTD with $\theta =$
Top: A design with an indifference interval $\theta \pm 0$.
Curve 1 (left), but may select a dose among levels 3, 4, and 5 un
A design with an indifference interval $\theta \pm 0$.
under both curves. The dark horizontal line in each plot indicates the target rate $\theta = 0$.
dotted lines indicate the indifference interval $\theta \pm \delta$.

5.3.2 Consistency Conditions of the CRM

Shen and O'Quigley [94] study the conditions under which the CRM is strongly consistent. It is useful to bring up two point are dealing with an asymptotic property, the conditions do n design in a two-stage CRM. Second, in this chapter, we will es of the Bayesian CRM under Conditions 4.1–4.8 stated in Secti be sketched in Section 5.4. As the initial design is irreleva consistency, so is the prior distribution $G(\beta)$ on the model parameter β $G(\beta)$ is not a degenerated distribution.

5.3.2.1 Home Sets

Now recall that $F_k(\beta) = F(d_k, \beta)$ for $k = 1, \ldots, K$ that $F_k(\beta)$ is monotone decreasing in β for all k dose level j

$$B_j = \{\beta \in [\underline{b}, \overline{b}] : |F_j(\beta) - \theta)| < |F_k(\beta) - \theta| \text{ for } k \neq j\},$$

where $\underline{b}$ and $\overline{b}$ are constants so that $F_1(\underline{b}) > \theta$ and $F_K(\overline{b}) < \theta$, and that $[\underline{b}, \overline{b}]$ compact subset of the parameter space of β (Condition 4.5). We call B_j the for dose level j, because the CRM will select dose level j if $\hat{\beta}_n \in B_j$ two conditions on the model $F_k(\beta)$ defined by $B_1 = [\underline{b}, b_2), B_j = (b_j, b_{j+1})$ for $j = 2, \ldots, K-1$, and $B_K = (b_K, \overline{b}]$ b_j solves

$$F_{j-1}(b_j) + F_j(b_j) = 2\theta \text{ for } j = 2, \ldots, K$$

Example 5.1. Consider the ψ-representation of the logistic model

$$F_k(\beta) = \frac{\exp[3 + \exp(\beta)\{\text{logit}(p_{0k}) - 3\}]}{1 + \exp[3 + \exp(\beta)\{\text{logit}(p_{0k}) - 3\}]}$$

with skeleton $p_{01} = 0.05, p_{02} = 0.12, p_{03} = 0.25, p_{04} = 0.40$, and $p_{05} = 0.$ trial with target $\theta = 0.25$, if we set $\underline{b} = -5.000$ and $\overline{b} = 5$.

$$B_1 = [-5.000, -0.280), B_2 = (-0.280, -0.093), B_3 = (-0.093, 0.097),$$
$$B_4 = (0.097, 0.288), \text{ and } B_5 = (0.288, 5.00].$$

5.3.2.2 Least False Parameters

Next, define

$$\beta_j = F_j^{-1}(p_j) \text{ for } j = 1, \ldots, K.$$

The parameter β_j may be called a *least false parameter* for dose level j the parameter value with which the model F_j at dose level j probability p_j. In the special case where all least false parameters are ide is, $\beta_j \equiv \beta_0$ for some β_0, the model $F(x, \beta)$ toxicity relationship and β_0 β_0 may not exist when the model is misspecified. However, the lea always exist, because F_j^{-1} exists by monotonicity of F_j (Condition 4.4).

5.3.2.3 Main Result

In Chapter 4, Section 4.3 lists the regularity conditions (C
CRM model. These assumptions are mild and verifiable. For the
tent, we require another assumption:

Condition 5.1 (consistency condition). *$\beta_j \in B_l$ for all j and some l.*

Theorem 5.3 (consistency). *Suppose that the dose–toxicity model $F(x, \beta)$
the regularity conditions given in Section 4.3. If Conditio
select dose d_l eventually with probability one, and $l = \nu$ is the true MTD.*

When $F(x, \beta)$ is a correct specification of the dose–toxicity curve, Condi
is satisfied. That is, the CRM is consistent when we use a corre
in general, consistency of the CRM requires that the least false parameters β_j
sufficiently close to each other so that they belong to the same home set.

Table 5.1 gives the least false parameter values β_j
under four dose–toxicity curves, where $\nu = 4$ for a target $\theta = 0.$
a correct specification of Curve 0, indicated by the fact that $\beta_j \equiv 0.18$ for all
a consequence, Condition 5.1 is met with $l = 4$, and $0.18 \in B_4$; the home sets
of model (5.6) are given in (5.7) in Example 5.1. Theorem 5.3 g
using model (5.6) will consistently select the true MTD unde
consistency to hold, Condition 5.1 requires only that $\beta_j \in B_4$
this condition under Curve 1 in Table 5.1, but not under Curves 2 and 3.

Table 5.1 Least false parameters β_j

j	Curve 0		Curve 1		Curve 2		Curve 3	
	p_j	β_j	p_j	β_j	p_j	β_j	p_j	β_j
1	0.02	0.18	0.01	0.21	0.03	0.09	0.01	0.30
2	0.05	0.18	0.05	0.18	0.11	0.01	0.02	0.35
3	0.13	0.18	0.12	0.19	0.21	0.05	0.09	0.25
4	0.25	0.18	0.25	0.18	0.25	0.18	0.25	0.18
5	0.41	0.18	0.46	0.12	0.28	0.34	0.54	0.02

As illustrated in the above computations, the least false pa
depend on the underlying unknown dose–toxicity curve. Ther
regularity conditions, Condition 5.1 cannot be directly us
between models. We will shortly return to this issue in Section 5.3.3.

5.3.2.4 A Relaxed Condition

When the underlying dose–toxicity curve is steep, the MTD is
neighboring doses on the probability scale; cf., Curve 3 in T
a reasonable dose finding method will likely identify the MTD
under a steep dose–toxicity curve. In other words, if a CRM de

MTD estimate under Curve 1 in Table 5.1, we expect that the sam
be consistent under Curve 3. Cheung and Chappell [21] thus po
condition as a sufficient condition for consistency:

Condition 5.2 (steepness condition).

$$\beta_k \in \bigcup_{j=k+1}^{K} B_j \; for\; k < l,\; \beta_l \in B_l,\; and\; \beta_k \in \bigcup_{j=1}^{k-1} B_j \; for\; k > l.$$

Condition 5.1 is stronger than Condition 5.2, and implies th
steepness condition to model (5.6) in Example 5.1 for $l = 4$, we obtain

$$\beta_1 \in (-0.280, 5.000], \beta_2 \in (-0.093, 5.000], \beta_3 = (0.097, 5.000],$$

$$\beta_4 \in (0.097, 0.288), \text{ and } \beta_5 \in [-5.000, 0.288)$$

From Table 5.1, we see that model (5.6) satisfies the steepnes
Curve 3, and hence anticipate that the CRM will converge to th
dose–toxicity curve. Figure 5.4 plots the probability of co
a two-stage CRM using model (5.6) under Curves 1–3 in Table 5.
since Condition 5.2 is met under Curve 3 (as well as Curve 1), t
correct MTD increasingly often as sample size increases.

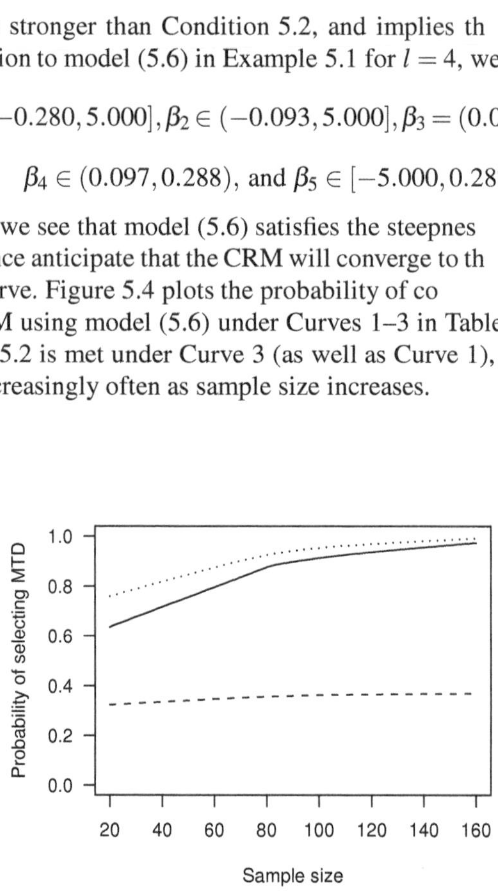

Figure 5.4 Probability of selecting the MTD (dose level 4) ve
CRM under dose–toxicity Curve 1 (solid), Curve 2 (dashed), a
The maximum likelihood CRM is used with model (5.6) and a grou

Curve 2 is not steep enough for model (5.6) as the steepness co
under Curve 2: $\beta_5 = 0.34 \notin [-5.000, 0.288)$
of selecting the MTD plateaus at about 40% under Curve 2 even a

grows. Thus, Condition 5.2 successfully predicts the failu
dose–toxicity scenario. More important, we observe that th
predictive of the CRM's poor performance at small samples. I
relevance of asymptotics to dose finding trials with small-t

5.3.3 *Model Sensitivity of the CRM*

Ideally, Conditions 5.1 and 5.2 would allow us to specify the model $F_k(\beta)$
the CRM is consistent. Unfortunately, both conditions requ
known dose–toxicity curve, and cannot be used in practice. T
verse approach that enumerates the set of dose–toxicity cur
model $F_k(\beta)$ will satisfies Condition 5.2 [21]. To illustrate, consider m
Example 5.1. Condition 5.2 for this model with $\nu =$
alent to

$$p_1 \in [0.00, 0.18), p_2 \in (0.00, 0.18], p_3 = (0.00, 0.18],$$

$$p_4 \in (0.18, 0.32), \text{ and } p_5 \in (0.32, 0.95]$$

by converting the intervals onto the probability scale via $p_k = F_k(\beta_k)$
the use of model (5.6) will lead to consistent MTD estimation
toxicity curve with $\nu = 4$, $p_1, p_2, p_3 < 0.18$ and $p_5 > 0.$
trated in Figure 5.5d. Condition 5.2 is satisfied if the dose–
pletely in the shaded region. In particular, Curves 1 and 3 in
in the shaded region, whereas Curve 2 lies outside the region
ing this CRM design is not consistent under Curve 2.

As the true dose–toxicity curve is unknown, so is the true MTD ν
need to repeat the calculation analogous to that of (5.9) for each possible ν
plots the regions of dose–toxicity curves under which the CR
each possible ν. We observe that the regions may cover curves that are not mon
increasing, but require that the neighboring doses of the true ν
ν in terms of the toxicity probability. The quantity of such di
as a numerical measure for the sensitivity of model (5.6), as follows.

Suppose again that $\nu = 4$ with $p_4 \in (0.18, 0.32)$. If p_3 is close to $\theta = 0.$
not as close as p_4, that is, $p_3 \in (0.18, p_4)$ such that $|p_3 - \theta| > |p_4 - \theta|$
(5.9) is violated and the CRM may converge to dose level 3. By t
$p_5 \in (p_4, 0.32]$
CRM using model (5.6) may converge to a dose with toxicity pro
the interval $(0.18, 0.32]$ if the true $\nu = 4$. The interval $(0.18, 0.32]$
indifference interval of this CRM design given $\nu = 4$.

Generally, for a trial with K test doses, for a given model $F_k(\beta)$
interval (θ_L, θ_U) of the CRM can be computed as follows: let

$$(\theta_{L,\nu}, \theta_{U,\nu}) = \begin{cases} (0, F_2(b_2)) & \text{if } \nu = 1 \\ (F_{\nu-1}(b_\nu), F_{\nu+1}(b_{\nu+1})) & \text{if } \nu = 2, \ldots, K-1 \\ (F_{K-1}(b_K), 1) & \text{if } \nu = K \end{cases}$$

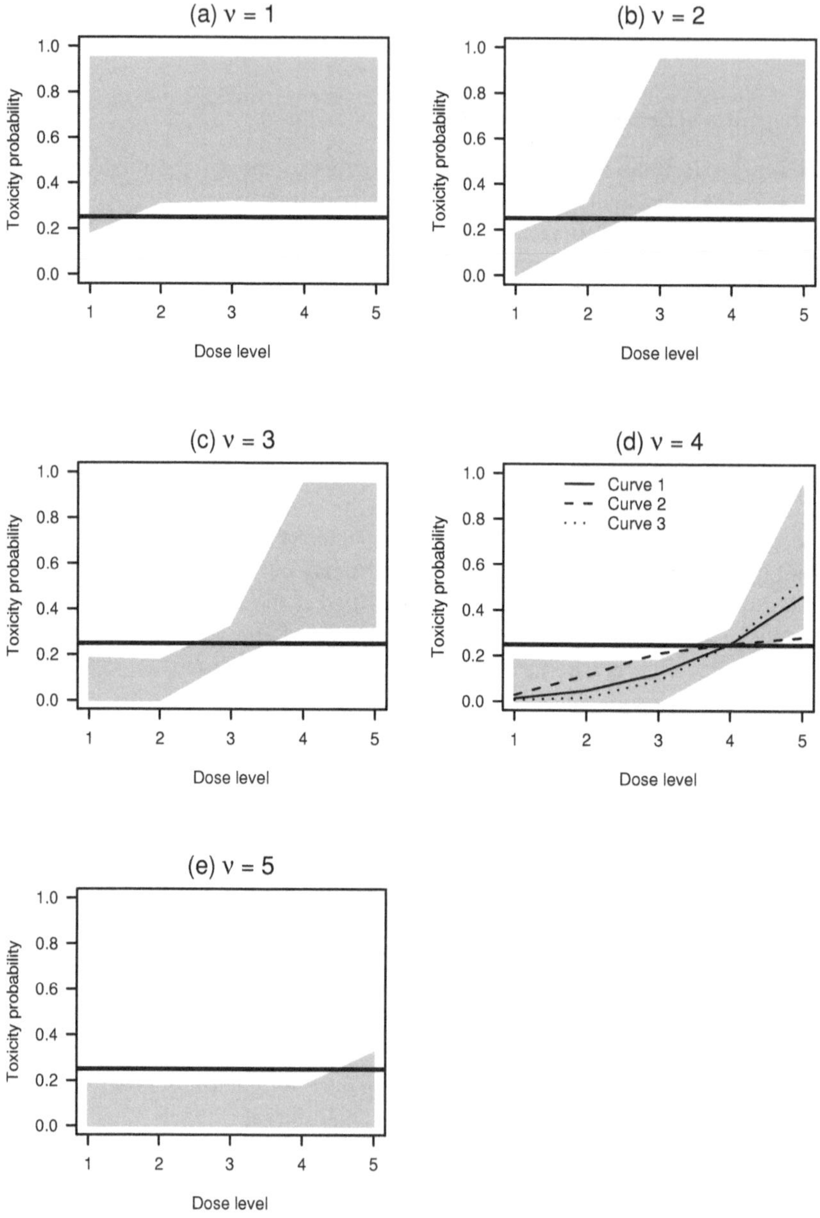

Figure 5.5 Condition 5.2 for logistic model (5.6) in Example 5.1 with $\theta = 0$. satisfies Condition 5.2 if the dose–toxicity curve lies completely in the shaded region.

where $b_2, \ldots, b_K$ solve (5.5). Then the overall indifference interval is given by min$\{\theta_{L,\nu} : \nu = 2, \ldots, K\}$ and $\theta_U = \max\{\theta_{U,\nu} : \nu = 1, \ldots, K-1\}$.

Table 5.2a displays the indifference interval of model (5.6) for each ν
table, we can deduce that this CRM model will eventually reco
toxicity probability somewhere between $(0.175, 0.325) = 0.250 \pm 0.$
that the CRM may select a dose outside this range if all the tes
too low or too toxic. This is apparently an issue about how dos
experimentation, rather than an issue of model sensitivity.

Table 5.2 Indifference intervals of three CRM models for $\theta = 0.25$

ν	(a) Model (5.6)		(b) Model (5.10)		(c) Model (5.17)	
	$\theta_{L,\nu}$	$\theta_{U,\nu}$	$\theta_{L,\nu}$	$\theta_{U,\nu}$	$\theta_{L,\nu}$	$\theta_{U,\nu}$
1	0.000	0.316	0.000	0.301	0.000	0.313
2	0.184	0.325	0.199	0.339	0.187	0.305
3	0.175	0.320	0.161	0.342	0.195	0.297
4	0.180	0.324	0.158	0.427	0.203	0.297
5	0.176	1.000	0.073	1.000	0.203	1.000
Overall	0.175	0.325	0.073	0.427	0.203	0.313

Consider another logistic CRM model:

$$F_k(\beta) = \frac{\exp\left[3 + \exp(\beta)\left\{\text{logit}(p'_{0k}) - 3\right\}\right]}{1 + \exp\left[3 + \exp(\beta)\left\{\text{logit}(p'_{0k}) - 3\right\}\right]}$$

with skeleton $p'_{01} = 0.05, p'_{02} = 0.10, p'_{03} = 0.25, p'_{04} = 0.45$, and $p'_{05} = 0.$
trial with $\theta = 0.25$. Table 5.2b shows the indifference intervals of this mode
particular, if the true $\nu =$
from a dose with toxicity probability as low as 0.073. Also, t
wider indifference intervals than model (5.6). These inter
numerical measures for model sensitivity and the basis for c
CRM models.

Another important point is that the model sensitivity in the
the skeleton, generally to a greater extent than the functional form $F(\cdot, \beta)$
(5.6) and (5.10) have the same logistic form but their indiff
a result of the choices of $\{p_{0k}\}$ and $\{p'_{0k}\}$
will serve as the theoretical basis of how the skeleton is calibrated in Chapter 8.

5.3.4 *Computing Model Sensitivity in R*

Model sensitivity of the CRM is evaluated by the function `crmsens`
package. The function requires input on the skeleton $\{p_{0k}\}$, the target θ
toxicity model (empiric or logistic), and an intercept for t
applies. The following R code evaluates the sensitivity of model (5.6).

```
> ### Evaluate sensitivity of a CRM model
```

```
> p0 <- c(0.05, 0.12, 0.25, 0.40, 0.55)
> theta <- 0.25
> foo <- crmsens(p0, theta, model="logistic", intcpt=3)
> foo$H          # home sets
            [,1]        [,2]
[1,] -5.00000000 -0.28046846
[2,] -0.28046846 -0.09340689
[3,] -0.09340689  0.09723351
[4,]  0.09723351  0.28842988
[5,]  0.28842988  5.00000000
> foo$iint      # indifference intervals
          [,1]      [,2]
[1,]        NA 0.3161698
[2,] 0.1838308 0.3245127
[3,] 0.1754882 0.3201194
[4,] 0.1798811 0.3240478
[5,] 0.1759521        NA
>
```

5.4 Proofs

5.4.1 *Coherence of One-Stage CRM*

Coherence of the one-stage Bayesian CRM requires Condition
following, assume without loss of generality that $F_k(\beta)$
for all k.

To establish coherence in escalation, we need to prove $x_{i+1} \leq x_i$ if $Y_i =$
all i. Under Condition 4.1, it suffices to show that $Y_i = 1$ implies $\hat{\beta}_i \leq \hat{\beta}_{i-1}$
$F_k(\hat{\beta}_i) \geq F_k(\hat{\beta}_{i-1})$ for all k
[16], who proves coherence for an extended class of CRM. First,

$$\begin{aligned}
\hat{\beta}_i - \hat{\beta}_{i-1} &= \frac{\int \beta L_i(\beta) dG(\beta)}{\int L_i(\beta) dG(\beta)} - \frac{\int \beta L_{i-1}(\beta) dG(\beta)}{\int L_{i-1}(\beta) dG(\beta)} \\
&= \frac{\int \beta L_i(\beta) dG(\beta) \int L_{i-1}(\beta) dG(\beta) - \int \beta L_{i-1}(\beta) dG(\beta) \int L_i(\beta) dG(\beta)}{\int L_i(\beta) dG(\beta) \int L_{i-1}(\beta) dG(\beta)}.
\end{aligned}$$

Because $L_i(\beta) = F(x_i, \beta) L_{i-1}(\beta)$ for $Y_i =$
can be rewritten as

$$\begin{aligned}
A &= \int \beta L_{i-1}(\beta) L_{i-1}(\phi) \{F(x_i, \beta) - F(x_i, \phi)\} dG(\beta) dG(\phi) \\
&= \int \phi L_{i-1}(\phi) L_{i-1}(\beta) \{F(x_i, \phi) - F(x_i, \beta)\} dG(\phi) dG(\beta)
\end{aligned}$$

The expression (5.12) is obtained by symmetry. Adding (5.11) and (5.12) gives

$$2A = \int L_{i-1}(\beta) L_{i-1}(\phi) (\beta - \phi) \{F(x_i, \beta) - F(x_i, \phi)\} dG(\phi) dG(\beta) < 0$$

because $(\beta - \phi)\{F(x_i, \beta) - F(x_i, \phi)\} < 0$ under Condition 4.2. Hence, $\hat{\beta}_i - \hat{\beta}_{i-1}$ when $Y_i = 1$.

Coherence of the one-stage Bayesian CRM is completed by anal for the case $Y_i = 0$.

5.4.2 *Consistency of the CRM*

Shen and O'Quigley [94] prove the consistency of the maximum under Condition 5.1, in addition to Conditions 4.1, 4.2, and Section 4.3. Readers are referred to their article for the details of the proof.

In this subsection, we sketch the proof of the consistency of the *Bayesian* using posterior mean under an additional assumption that Co A key result in the proof of the maximum likelihood CRM is that likelihood estimate $\tilde{\beta}_n$, given the observation history H_{n+1} B_ν almost surely when the sample size n that $\tilde{\beta}_n \in B_\nu$ for any possible history H_{n+1} assignment mechanism.

Let $l_n(\beta) = \log L_n(\beta)$ denote the log-likelihood of β tives, respectively, denoted by $l_n'(\beta)$ and $l_n''(\beta)$. Precisely,

$$l_n'(\beta) = \sum_{k=1} z_k \frac{F_k'(\beta)}{F_k(\beta)} + (n_k - z_k) \frac{-F_k'(\beta)}{1 - F_k(\beta)}$$

where z_k and n_k treated at dose k. Since $l_n'(\tilde{\beta}_n) = 0$, we can apply Taylor's expansion and obtain

$$l_n(\beta) = l_n(\tilde{\beta}) + \frac{1}{2}(\beta - \tilde{\beta})^2 l_n''(\tilde{\beta}_n^*)$$

for some $\tilde{\beta}_n^*$ between β and $\tilde{\beta}_n$, where $l_n''(\beta) =$

$$\sum_{k=1}^{K} z_k \left[\frac{F_k(\beta)F_k''(\beta) - \{F_k'(\beta)\}^2}{\{F_k(\beta)\}^2} \right] + (n_k - z_k) \left[\frac{-F_k''(\beta) + F_k(\beta)F_k''(\beta) - \{F_k'(\beta}{\{1 - F_k(\beta)\}^2} \right.$$

Under Condition 4.8, we have

$$\bar{l}_n'' \equiv \sup_{\beta \in [\underline{b}, \bar{b}]} l_n''(\beta) \to -\infty \text{ as } n \to \infty$$

almost surely. Now, we express the posterior mean of β given history H_{n+1} as

$$\hat{\beta}_n = \frac{\int_{-\infty}^{\infty} \beta L_n(\beta)\, dG(\beta)}{\int_{-\infty}^{\infty} L_n(\beta)\, dG(\beta)} = \tilde{\beta}_n + \frac{\int_{-\infty}^{\infty} (\beta - \tilde{\beta}_n) L_n(\beta)\, dG(\beta)}{\int_{-\infty}^{\infty} L_n(\beta)\, dG(\beta)}.$$

Furthermore,

$$\left| \int_{-\infty}^{\infty} (\beta - \tilde{\beta}_n) L_n(\beta)\, dG(\beta) \right|$$

$$= \left| \int_{-\infty}^{\infty} (\beta - \tilde{\beta}_n) \exp\left\{ l_n(\tilde{\beta}_n) + \frac{1}{2}(\beta - \tilde{\beta}_n)^2 l_n''(\tilde{\beta}_n^*) \right\} dG(\beta) \right|$$

$$\leq L_n(\tilde{\beta}_n) \int_{-\infty}^{\infty} \left| \beta - \tilde{\beta}_n \right| \exp\left\{ \frac{1}{2}(\beta - \tilde{\beta}_n)^2 \bar{l}_n'' \right\} dG(\beta)$$

The above integral can be further broken into

$$\int_{\tilde{\beta}_n - \varepsilon}^{\tilde{\beta}_n + \varepsilon} \left| \beta - \tilde{\beta}_n \right| \exp\left\{ \frac{1}{2}(\beta - \tilde{\beta}_n)^2 \bar{l}_n'' \right\} dG(\beta) \leq \varepsilon$$

for a given $\varepsilon > 0$, and

$$\left(\int_{-\infty}^{\tilde{\beta}_n - \varepsilon} + \int_{\tilde{\beta}_n + \varepsilon}^{\infty} \right) \left| \beta - \tilde{\beta}_n \right| \exp\left\{ \frac{1}{2}(\beta - \tilde{\beta}_n)^2 \bar{l}_n'' \right\} dG(\beta) \rightarrow$$

as $n \rightarrow \infty$ by (5.13). Since we can choose ε
dently of n, the second term in (5.14) is arbitrarily close to 0 as sample
We have thus shown

$$|\hat{\beta}_n - \tilde{\beta}_n| \rightarrow 0$$

almost surely. Since $\tilde{\beta}_n \in B_\nu$
same conclusion for $\hat{\beta}_n$ because B_ν is an open set.

5.5 Exercises and Further Results

Exercise 5.1. Using Definition (5.4) of the home sets, convince yourself that $\hat{\beta}_n$
implies the Bayesian CRM will recommend dose k for the next patient.

Exercise 5.2. Verify that (5.7) in Example 5.1 solves (5.5).

Exercise 5.3. Verify the computation of the indifference intervals in Tab
the following logistic CRM model for a trial with $\theta = 0.25$:

$$F_k(\beta) = \frac{\exp\left[3 + \exp(\beta)\left\{\text{logit}(p_{0k}'') - 3\right\}\right]}{1 + \exp\left[3 + \exp(\beta)\left\{\text{logit}(p_{0k}'') - 3\right\}\right]}$$

with skeleton $p_{01}'' = 0.07, p_{02}'' = 0.15, p_{03}'' = 0.25, p_{04}'' = 0.35$, and $p_{05}'' = 0.45$.

Exercise 5.4. Using similar arguments developed in Section 5.4, prove tha
Bayesian CRM with the posterior mode is coherent (cf. Exercise 4.4).

Chapter 6

Empirical Properties

6.1 Introduction

The two main operating characteristics of a dose finding desi
distribution of the selected dose and the in-trial dose allo
accuracy of the design, and the latter safety. This chapter i
indices for accuracy and safety in Section 6.2; these indice
design comparison. In Section 6.3, we present a nonparametr
for a method's performance.

6.2 Operating Characteristics

6.2.1 Accuracy Index

Table 6.1 compares the distribution of the selected dose of a
two-stage CRM using models (5.6), (5.10), and (5.17) for a trial with $\theta = 0.$
sample size is $N =$
ability of correct selection (PCS) of the true MTD is indicat
The PCS is the most immediate index for accuracy, which can be
different designs. For example, the design using model (5.1
Curves 1–3 in Table 6.1, but worst under Curve 4. In addition,
line with asymptotic reasoning: the accuracy improves as th
comes steep; the poor performance of model (5.10) under Curv
by its violation of Condition 5.2. For a similar asymptotic r
have low PCS under Curve 2.

On the other hand, the entire distribution of selected dose d
detailed information than what the PCS alone suggests. For e
importance to look at the probability of selecting an overly
dose level 5 under Curves 1 and 3 in Table 6.1. This having been
comparison purposes, it is useful to have a single numerical
the entire distribution and reflects a design's accuracy. In
following accuracy index:

$$\mathscr{A}_N = 1 - K \times \frac{\sum_{k=1}^{K} \rho_k \times \text{Probability of selecting dose } k}{\sum_{k=1}^{K} \rho_k}.$$

where ρ_k is a discrepancy measure between the true toxicity probability p_k

Table 6.1 Distribution of the selected dose and the accuracy for a trial with a target $\theta = 0.25$ and $N = 20$

Model	Probability of selecting dose					Accuracy index $\mathscr{A}_{20}$			
	1	2	3	4	5	sq	abs	0-1	od[a]
Curve 1:	.01	.05	.12	**.25**	.46				
(5.6)	.00	.02	.23	**.55**	.19	.58	.51	.44	.28
(5.10)	.00	.02	.24	**.63**	.11	.69	.61	.53	.51
(5.17)	.00	.02	.24	**.51**	.24	.52	.45	.38	.16
Curve 2:	.03	.11	.21	**.25**	.28				
(5.6)	.01	.13	.33	**.26**	.28	.75	.52	.07	.34
(5.10)	.01	.13	.36	**.33**	.17	.75	.54	.16	.44
(5.17)	.01	.12	.27	**.25**	.37	.77	.54	.06	.27
Curve 3:	.01	.02	.09	**.25**	.54				
(5.6)	.00	.00	.20	**.63**	.16	.58	.56	.54	.36
(5.10)	.00	.00	.17	**.72**	.11	.70	.67	.65	.55
(5.17)	.00	.01	.21	**.59**	.20	.51	.49	.48	.25
Curve 4:	.03	.05	.08	.12	**.25**				
(5.6)	.00	.01	.07	.25	**.66**	.74	.66	.57	.66
(5.10)	.00	.01	.09	.41	**.48**	.62	.50	.35	.50
(5.17)	.00	.01	.05	.20	**.74**	.81	.75	.68	.75

Note: Numbers associated with the MTD are in bold.
[a]*The overdose error is used with* $\alpha = 0.20$.

level k and the target θ
selected dose through its weighted average with weights $\rho_j / \sum_k \rho_k$
index indicates high accuracy, and the maximum value of the i
make a blind guess in the absence of any observation, that is, $N =$
selecting one of the K test doses with equal probability $1/K$
that $\mathscr{A}_0 = 0$. Therefore, the index (6.1) measures the information gain
use of the observations. Finally, the index $\mathscr{A}_N$
improper use of data could do more harm than good.

Among many possible choices of ρ_k

1. Squared discrepancy (sq): $\rho_k = (p_k - \theta)^2$.
2. Absolute discrepancy (abs): $\rho_k = |p_k - \theta|$.
3. Classification error (0-1): $\rho_k = I(k \neq \nu)$, where $I(\cdot)$ is an indicator function.
4. Overdose error (od): $\rho_k = \alpha(\theta - p_k)^+ + (1-\alpha)(p_k - \theta)^+$ for some $\alpha \leq 0$.

The discrepancy measure ρ_k
Precisely, the denominator $\sum_k \rho_k$
from the target θ
may commit under the dose–toxicity curve. Consider Curve 3 i
use the absolute discrepancy error, then $\rho_1 = 0.24, \rho_2 = 0.23, \rho_3 = 0.16, \rho_4 = 0$

$\rho_5 = 0.29$ and $\sum_k \rho_k = 0.$
than Curve 3, has $\sum_k \rho_k = 0.78$; Curve 2, the flattest of all three, has $\sum_k \rho_k =$
Intuitively, it is easy to see that selecting dose level 5 und
0.28 toxicity probability) is not as serious a mistake as sel
Curve 3 (which has a 0.54 toxicity probability).

Table 6.1 gives the accuracy index $\mathscr{A}_{20}$
discrepancy meaures. Most of the time, using different ρ_k
rankings of the designs. For example, the CRM with model (5.1
and 3 by all discrepancy measures, and model (5.17) is best fo
hand, under Curve 2, model (5.10) is compared favorably to th
the classification error or overdose error ($\alpha = 0.$
selecting dose 5 under Curve 2 is not a serious mistake. Howev
to-one correspondence between the classification error and
measure ignores the insignificance of the error and penalize
relatively severely for choosing dose 5 more often than mode
the overdose error with $\alpha = 0.$
mistake. In contrast, the absolute error appears quite sens
similar accuracy scores under Curve 2, while the frequent se
models (5.6) and (5.17) is appropriately penalized under Cu
error depicts a similar picture.

6.2.2 *Overdose Number*

The overdose (OD) number, defined as the average number of pat
dose above the true MTD per trial, is a safety index of a dose fin
trial dose allocation by three CRM designs used in Table 6.1 i
Under Curves 1–3, the OD number is the average number of assig
level 5; under Curve 4, the OD number is invariably zero becau
As a benchmark, a balanced randomization scheme on average a
each dose for a trial with $K = 5$ dose levels and sample size $N =$
we expect an OD number of 4 patients under Curves 1–3 by balanc
The CRM using models (5.6) and (5.10) improve safety by havin
lower than 4 under Curves 1–3. In contrast, using model (5.17
concerns as it could lead to a OD number higher than balanced r
model may not be used on this ground.

6.2.3 *Average Toxicity Number*

Another summary index for safety is the average toxicity num
which is roughly equal to

$$\sum_{k=1}^{K} p_k \times \text{Number of assignments at dose } k.$$

A large ATN certainly causes safety concerns. On the other ha
necessarily an accurate indication of safety. For example,

Table 6.2 In-trial dose allocation and the average toxicity used in Table 6.1

Model	Number allocated to dose					ATN
	1	2	3	4	5	
Curve 1:	.01	.05	.12	.25	**.46**	
(5.6)	2.3	3.0	5.0	6.5	**3.3**	4.0
(5.10)	2.2	2.8	5.7	7.2	**2.1**	3.7
(5.17)	2.3	2.7	4.7	6.2	**4.2**	4.3
Curve 2:	.03	.11	.21	.25	**.28**	
(5.6)	2.9	4.6	5.4	3.9	**3.2**	3.7
(5.10)	2.8	4.5	6.3	4.2	**2.2**	3.6
(5.17)	2.9	4.1	4.8	3.8	**4.4**	3.8
Curve 3:	.01	.02	.09	.25	**.54**	
(5.6)	2.1	2.4	4.7	7.5	**3.3**	4.2
(5.10)	2.1	2.3	5.0	8.3	**2.3**	3.8
(5.17)	2.1	2.3	4.5	7.0	**4.0**	4.4
Curve 4:	.03	.05	.08	.12	.25	
(5.6)	2.4	2.8	3.3	4.5	7.0	2.8
(5.10)	2.4	2.7	3.9	5.7	5.3	2.6
(5.17)	2.4	2.5	2.9	4.1	8.1	3.0

Note: Numbers associated with OD are in bold.

highest dose is the MTD, using model (5.17) causes the larges this simply reflects that it treats more patients at the MTD th Therefore, a high ATN is desirable in this scenario. To set up the ideal situation where all patients are being treated at t expect $N\theta$ toxic outcomes per trial. Therefore, a design that yields roughly $N\theta$ outcomes per trial can be considered safe, as long as the OD nu relative to balanced randomization.

The OD number and the ATN extract complementary safety infor the in-trial allocation of a design. In Table 6.3, Design A is all patients treated at the true MTD. If we use ATN as the only s design B will be declared as safe as design A although it treat overdose. In contrast, if we use OD number only, design C appe indeed overconservative by treating the majority at an unde is indicated by a low ATN.

6.3 A Nonparametric Optimal Benchmark

In a trial, each patient is given a dose, with the toxicity out dose. Only a *partial* receives dose level 3 in a trial and has a nontoxic outcome. We monotonicity that he would have had a nontoxic outcome at dos

Table 6.3 In-trial allocation of three generic designs for a trial with $\theta = 0.25$, $K = N = 20$ under a curve with $p_1 = 0.01, p_2 = 0.25$, and $p_3 = 0.50$

Design	Number allocated to dose			ATN
	1	2	3	
p_k:	.01	.25	**.50**	
A	0	20	**0**	5
B	10	0	**10**	5
C	15	5	**0**	1.4

Note: Numbers associated with OD are in bold.

we will have no information whether the patient would have su
had he been given a dose that is above dose level 3.

In a computer-simulated trial, however, by generating the t
each patient, we can observe the toxicity outcome at every test dose, that is, a *plete* toxicity profile. Table 6.4 gives the complete toxicity profi
patients under Curve 1 in Table 6.1. Recall that u_i
of patient i, so that the patient will have a toxic outcome if treated at dose k
$u_i \leq p_k$. Take patient 9, for example. He will have a nontoxic outcome
and 3 because $u_9 = 0.175 > p_k$ for $k = 1, 2,$
because $u_9 < p_4, p_5$.

Since the toxicity tolerance u_i
information in Table 6.4 is generally not available from a tr
have the complete toxicity profiles in a simulated trial, we c
probability p_k by the sample proportion $\hat{p}_k$
at each dose, and select the MTD in accordance with these esti
row in Table 6.4 gives the sample proportions of this simulat
we may choose dose level 4 as the MTD for a trial with a target $\theta = 0.$
is closest to θ. (Here, we adopt the conservative convention that when more
dose is equally close to θ, the largest dose with $\hat{p}_k \leq \theta$
this MTD estimate is optimal in the sense that $\hat{p}_k$ is unbiased for p_k
Cramér-Rao lower bound. This design may thus be called a non
design.

While the nonparametric optimal design cannot be implement
where complete toxicity profiles are not available, its perf
over many simulation runs and be used as a benchmark for effici
the accuracy indices $\mathscr{A}_{20}$
are 0.63, 0.45, 0.77, and 0.81, respectively, under Curves 1
optimal design is superior to the CRM under Curves 1, 3, and 4.
because the former uses information that is not available to
hand, some of the CRM designs get quite close to the optimal be
quite interesting to note that the CRM outperforms the nonpa
under Curve 2. While it appears to be impossible at first, this
fact that the CRM utilizes parametric dose–toxicity assump

Table 6.4 The complete toxicity profile of 20 simulated patie
A "0" indicates a nontoxic outcome at a dose, and "1" a toxic outcome

i	u_i	Toxicity probability p_k				
		.01	.05	.12	.25	.46
1	.571	0	0	0	0	0
2	.642	0	0	0	0	0
3	.466	0	0	0	0	0
4	.870	0	0	0	0	0
5	.634	0	0	0	0	0
6	.390	0	0	0	0	1
7	.524	0	0	0	0	0
8	.773	0	0	0	0	0
9	.175	0	0	0	1	1
10	.627	0	0	0	0	0
11	.321	0	0	0	0	1
12	.099	0	0	1	1	1
13	.383	0	0	0	0	1
14	.995	0	0	0	0	0
15	.628	0	0	0	0	0
16	.346	0	0	0	0	1
17	.919	0	0	0	0	0
18	.022	0	1	1	1	1
19	.647	0	0	0	0	0
20	.469	0	0	0	0	0
Sample proportion $\hat{p}_k$		.00	.05	.10	.15	.35

the *nonparametric* optimal design. It indicates the potential advantage of the that allows borrowing strength from information across dos toxicity model. O'Quiley et al. [77] give additional numeri is indeed near the optimal benchmark.

6.4 Exercises and Further Results

Exercise 6.1. Prove or verify the following statements:

a. There is a one-to-one correspondence between the PCS and t using the classification error.

b. The absolute discrepancy measure is a special case of the overdose error.

Part II

Design Calibration

Chapter 7

Specifications of a CRM Design

7.1 Introduction

A CRM design is completely specified by a set of design paramet
the set of *clinical parameters*
clinical objective, experimental setup, and practical con
parameters usually require inputs from clinical investiga
clinician-input parameters
θ, the number of test doses K, sample size N, the prior MTD ν_0
of the trial x_1.

The statistical component of a CRM design is specified by the
These parameters directly specify the dose–toxicity model
parameters include the functional form of the dose–toxicity function $F(\cdot,\beta)$
tial guesses p_{0k}s of toxicity probabilities associated with the test doses,
distribution $G(\beta)$ of the model parameter β
namely, when $x_1 < d_{\nu_0}$, one will also need to specify a dose sequence $\{x_{i,0}\}$
initial design.

This division is not the only way to classify the CRM design pa
to be used unalterably. While some design parameters clearl
(e.g., θ), other clinical parameters (e.g., N and ν_0
to the statistical properties of the design. Also, some mode
appear to require clinician inputs (e.g., p_{0k}
parameters provides a pragmatic system under which the stat
defined subsequent to the clinical parameters, so that the CR
good operating characteristics with respect to the clinica
following chapters (and this book) is the specification of th
although brief guidelines on the choice of the clinical para
in Section 7.2 of this chapter. Section 7.3 prescribes a road
model parameters and the initial design, and outlines the co
elaborated in the next three chapters. Section 7.4 presents
describes a trial-and-error approach used to choose the des
trials.

7.2 Specifying the Clinical Parameters

7.2.1 Target Rate θ

For chemotherapy trials in cancer patients, a DLT is typical
or higher toxicity according to the National Cancer Institu
Criteria for Adverse Events, and the convention for the target toxicity rate θ
somewhere between 0.20 and 0.30. While the rationale for thi
entirely clear apart from its rough agreement with the 3+3 al
outside this range should be adequately justified.

For other disease areas, the target rate θ
effects and the potential benefits to that patient populatio
example. Generally, it is useful to direct the clinical inve
their clinical experience by asking questions such as "what
effects of new drug?", "are these events expected to occur mo
higher dose?", "what is the toxicity rate of the standard tre
comparable patient population?" and "will a higher toxicit
the experimental treatment than with the standard treatment?"

7.2.2 Number of Test Doses K

The choice of test doses involves the determination of dose i
situations where prior information is available to identif
regimens, the specified dose increments will in turn determine the number K
doses for the trial.

There has been a perception that the dose increments are set b
on the modified Fibonacci sequence defined by the diminishing
$2.00, 1.67, 1.50, 1.33, 1.33, \ldots$
used [88]. In fact, there is no biological basis for the use of
Rather, the magnitude of increment is closely tied to the nat
cancer vaccine dose finding studies, each subsequent dose in
in terms of plague-forming units [51]. In drug trials, dose i
constant. Also, pharmacokinetic data in a few pilot patient
whether two doses are pharmacologically separated. While t
level tends to increase with dose, there is considerable var
ues. If two successive doses have comparable range of plasma
will likely demonstrate similar pharmacological properti
be differentiated in terms of therapeutic benefits. Thus, te
or the average of both, may prove sufficient.

7.2.3 Sample Size N

In general, we expect the required sample size N
doses increases. When planning a trial, we may make three pra
and considerations:

- The CRM will allocate the majority of the subjects roughly ev
  MTD and a neighbor dose.
- The other $K-2$ doses will receive about 3 subjects each.
- The number of subjects treated at the MTD should be greater than $1/\theta$
  will expect to observe at least one toxic observation at the MTD.

Based on these considerations, we suggest that the clinical
constrained so that

$$\frac{N-3(K-2)}{2} > \frac{1}{\theta}.$$

For example, if $N = 20$ subjects are available for a trial with a target $\theta =$
the inequality prescribes $K <$
test 6 or more dose levels. The criterion (7.1) has been shown
rule of thumb based on extensive simulations and experience
performance of the one-stage Bayesian CRM.

Alternatively, it may be natural to view the criterion (7.1)
sample size N for given θ and K. For example, for a trial with $\theta = 0.25$ and $K =$
doses, the inequality (7.1) implies $N >$
inequality is by no means sharp, and should not be used as a for
sample size. It may at first seem unusual to consider N
than a statistical parameter. In phase I trial practice, how
idea of the feasible range of sample size based on patient and
Therefore, it appears to be a reasonable strategy to start pl
N along with θ and K
screen out underresourced studies. In this regard, the ineq
as a sample size *constraint* instead of a sample size *formula*
constraint for the two-stage CRM can be found in Chapter 10.

After the model parameters and the initial design are specifi
should) reevaluate the required sample size based on the sta
design by simulations (See Figure 7.1).

7.2.4 Prior MTD ν_0 and Starting Dose x_1

The prior MTD ν_0
that is, the dose level given to the first patient enrolled in t
starting dose x_1 is a purely clinical decision based on the current state of kn
about the experimental drug. Ideally, the prior MTD ν_0
level, as originally prescribed by the CRM. However, it is no
clinical investigators have reasons to believe that the MTD
doses, and are not comfortable to start the trial at any dose t
conventional use of the drug in other patient populations, w
dose d_1. Should this be the case, a two-stage CRM will be an appropria
escalation strategy. A practical recommendation is to set ν_0

that is less than or equal to the median dose level. That is, set

$$\nu_0 = \begin{cases} K/2 & \text{for even } K \\ (K+1)/2 & \text{for odd } K \end{cases}$$

The convention (7.2) is based on the consideration of statis
CRM design; this point will be elaborated in Chapter 9. Accor
the prior MTD ν_0 is considered a model parameter rather than a clinical input
tuned to obtain good operating characteristics.

7.3 A Roadmap for Choosing the Statistical Component

As indicated by the number and dimension of the design parame
CRM design for a dose finding study is a complex process. Also,
performance depends on how these parameters are chosen, thi
for the success of a CRM application. This section outlines a
to obtain "good" choices for the statistical component, com
design parameters:

Functional form of dose–toxicity curve.
the functional form of the dose–toxicity curve $F(\cdot,\beta)$
choices of F, the two most commonly used functions in the CRM literature a
empiric function (3.2) and the logistic function (3.3) with a fixed intercept $a_0 =$
the consistency and coherence properties of the CRM do not de
choice of F as long as they satisfy the regularity conditions in Section
the empiric or the logistic functions shall prove sufficient
following chapters, we will see calibration techniques and
function with intercept $a_0 =$
for the empiric function in the literature [60].

Initial guesses of toxicity probabilities (skeleton).
model is the set of labels d_ks associated with the test doses. Choosing d_k
equivalent to specifying the initial guesses p_{0k}
with the doses, the so-called skeleton. The choice of the ske
pact on the operating characteristics of the CRM. Ideally, t
be chosen as clinical parameters so as to reflect the clinicia
test doses. Such information is seldom available in practic
pragmatic approach to determining a set of p_{0k}s that achieve a prespecified δ
of sensitivity (see Chapter 5, Section 5.3.1). This approac
process as it reduces the number of model parameters from K to 1. As δ
is an asymptotic concept, the choice of δ
cussed.

Prior distribution.
distribution for the model parameter β
and the logistic (3.3) functions, for which the parameter is
the real line, we can focus on any prior distribution that tak
Specifically, Chapter 9 will illustrate the calibration tec

that is,

$$\beta \sim N(\hat{\beta}_0, \sigma_\beta).$$

In light of the fact that the theoretical (asymptotic) prope particular form of the prior distribution, there are two pra normality choice. First, posterior computations using Gau under parameterization (3.2) and (3.3) are accurate when a n This is partly because of good normality approximation of th even when the sample size is small. Second, the Bayesian CRM u invariant to the mean of a prior that constitutes a location- allows us to set

$$\hat{\beta}_0 = 0$$

without loss of generality and further simplify the calibra completely specified by a single number, that is, the prior standard deviation σ_β

Initial design.
starting dose x_1 is lower than the model-based prior MTD d_{ν_0} to specify a dose sequence $\{x_{i,0}\}$ general characterization of dose sequences and discuss the calibration of $\{x_{i,0}$ particular, while a two-stage CRM is motivated by the inclin we will see that an overconservative initial design is incom objective θ. The initial design calibration will thus be guided by the co criterion (see Definition 5.1).

Many of the model choices made above, such as the use of logistic F prior, are motivated by the practical needs for simplicity. the CRM calibration process so that the study statistician w a reasonable design for a dose finding study in a *timely and reproducible* Figure 7.1 presents a diagrammatic scheme for the reduced sp the following design parameters:

- Half-width δ of the indifference interval (Chapter 8)
- Standard deviation σ_β of the normal prior distribution (Chapter 9)
- Initial dose sequence $x_{1,0}, x_{2,0}, \ldots, x_{N,0}$ (Chapter 10)

Each of these will be dealt with in detail in the next three chapters.

7.4 The Trial-and-Error Approach: Two Case Studies

7.4.1 The Bortezomib Trial

The bortezomib trial introduced in Chapter 1 evaluated $K =$ of bortezomib when administered in combination with the sta in lymphoma patients. The study used the standard definition endpoint, and the trial objective was to identify a dose associated with a $\theta =$ toxicity probability.

The study design was a one-stage Bayesian time-to-event CRM CRM. In this section, as we focus on the calibration of the dos

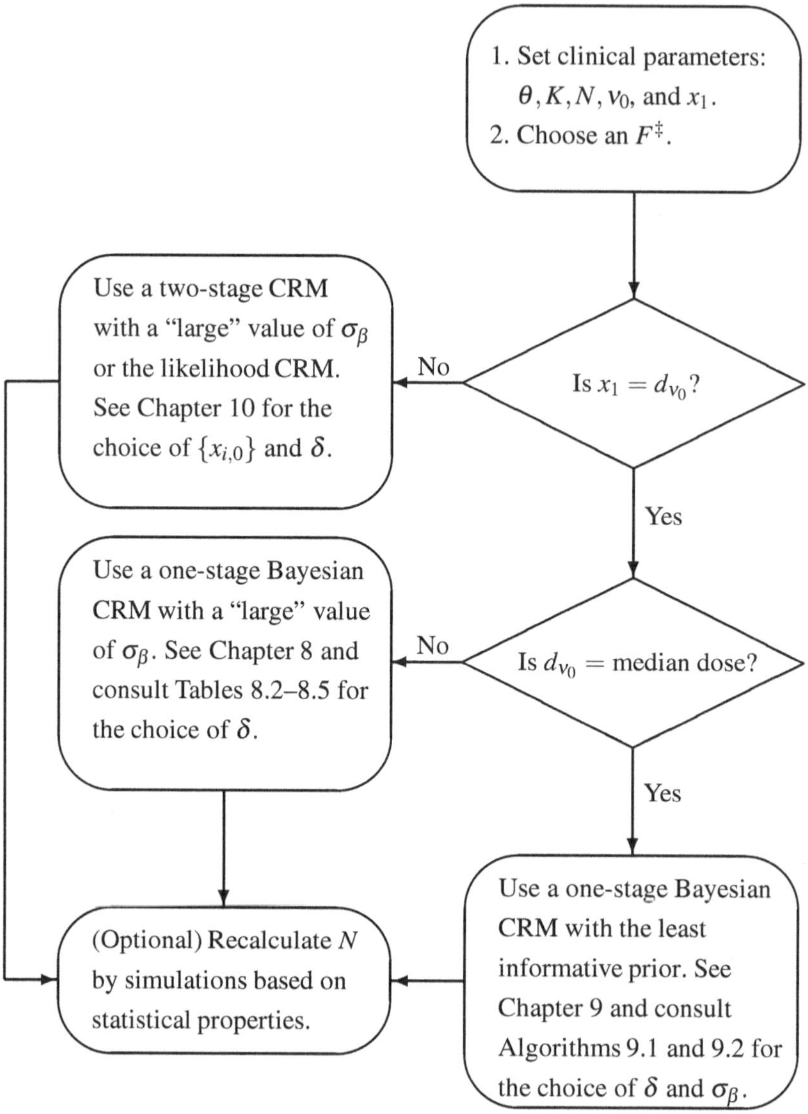

‡The subsequent chapters will use the logistic function (3.3) with $a_0 = 3$.

Figure 7.1 A flowchart for the specification of the model param
the CRM.

we will disregard the time-to-event component, for which we
ter 11. The starting dose was level 3, that is, $x_1 = d_3$
see Chapter 1, Table 1.1 for a detailed description of the dos
we specified the prior MTD $\nu_0 =$
was set at 18 (as a multiple of 3) based on budget and trial dura
the criterion (7.1) with $K = 5$ and $\theta = 0.25$, we got

$$N > \frac{2}{\theta} + 3(K - 2) = 17.$$

Thus, a sample size of 18 seemed to be reasonable. In the actua
patients were enrolled in the phase I portion, and one patien
study (see Figure 1.1).

At the planning stage of the study, we considered the empiric
a normal prior on β with mean 0 and variance $\sigma_\beta^2 = 1$.
choosing this model were convenience and convention: the em
the only dose–toxicity function we had programmed for at tha
distribution was determined based on an existing reference [80].

Next, we specified the initial guesses of DLT probabilities by $p_{03} = 0.$
the assumed $\nu_0 = 3$, along with $p_{01} = 0.05, p_{02} = 0.12, p_{04} = 0.40$, and $p_{05} =$
The overall indifference interval of this CRM model was $0.25 \pm 0.$
was made by trial and error: an initial skeleton was chosen wi
in the literature [25], with a constraint that the model's in
wider than 20 percentage points, that is, the half-width $\delta \leq 0.$
with this initial skeleton was evaluated by simulations und
scenarios given in Table 7.1. Then the skeleton was iterativ
the simulated operating characteristics. This process was
a haphazard manner, although at the end we felt the operating
final CRM model with $N =$
will be presented in Chapter 8.

Table 7.1 The simulation scenarios used in the bortezomib trial protocol

Scenario	p_1	p_2	p_3	p_4	p_5
1	**.25**	.40	.45	.55	.60
2	.05	**.25**	.40	.45	.55
3	.05	.05	**.25**	.45	.55
4	.05	.05	.08	**.25**	.45
5	.05	.05	.08	.12	**.25**

Note: Numbers associated with the MTD are in bold.

7.4.2 NeuSTART

The NeuSTART introduced in Example 2.1 studied the use of hig
acute ischemic stroke patients. In the trial, the primary sa

posttreatment elevation of peak liver or muscle enzyme leve
to identify the MTD associated with a $\theta = 0.$
reached in collaboration with the clinical investigators b
of 3% event rate in the untreated study population; a target rate θ
background was deemed acceptable after taking into account
the disability caused by an acute stroke could be severe, and
would likely be reversed by withdrawing treatment. In the tr
withdrawal would be counted as reaching the primary endpoint.

The NeuSTART used a two-stage time-to-event CRM for dose escalation at K
test doses. As in the previous subsection, we disregard the t
here. The dose schedules used in the trial are shown in Table 7
was specified by the empiric function (3.2) with skeleton $p_{01} = 0.02, p_{02} = 0.$
$p_{03} = 0.10, p_{04} = 0.18, p_{05} = 0.$
was 0.10 ± 0.04. The prior distribution was $\beta \sim N(0, 1.34)$
process of the NeuSTART was by and large similar to that of the
was characterized by ad hoc iteration of the skeleton. The pr
various sample sizes ranging from 24 to 36 (only multiples of
and the final sample size was set at $N = 33$.

Table 7.2 Dose schedules used in the NeuSTART and cohort size

Dose tier	Sample size	Lovastatin dose	
		0–72 hours	days 3–30
1	3	1 mg/kg/day	20 mg/day
2	3	3 mg/kg/day	20 mg/day
3	6	6 mg/kg/day	20 mg/day
4	9	8 mg/kg/day	20 mg/day
5	12	10 mg/kg/day	20 mg/day

This CRM model reflected the prior belief that the MTD was amon
doses by specifying $\nu_0 =$
deemed it appropriate to start the trial at the standard lova
(dose tier 1) and escalate through the five dose tiers sequent
initial design was defined by the dose sequence

$$x_{i,0}^{Neu} = \begin{cases} d_1 & \text{for } i = 1,2,3 \\ d_2 & \text{for } i = 4,5,6 \\ d_3 & \text{for } i = 7,\ldots,12 \\ d_4 & \text{for } i = 13,\ldots,21 \\ d_5 & \text{for } i = 22,\ldots,33 \end{cases}$$

and would be used to assign doses before any safety event occu
in the case where there was no event throughout the trial, thr
treated at dose level 1, three at level 2, and so on, so that it w
to reach the highest dose level. The determination of the dos

was also by trial and error. In fact, calibrating a two-stage
complicated than calibrating a one-stage CRM by this trial-
the process would jointly iterate the initial dose sequence $\{x_{i,0}\}$
p_{0k}s. However, at the end, the initial design satisfied two criteria:

1. The clinical investigators of the study thought that the escalation speed of was appropriately cautious.
2. The initial design was not overconservative in that it was provides details.

Furthermore, the model and the initial design that resulted found to perform well under the five dose–toxicity scenarios given in Table 7.3.

Table 7.3 The simulation scenarios used in the NeuSTART protocol

Scenario	p_1	p_2	p_3	p_4	p_5
1	**.10**	.25	.30	.35	.40
2	.04	**.10**	.25	.30	.35
3	.01	.04	**.10**	.25	.30
4	.01	.01	.04	**.10**	.25
5	.01	.01	.01	.04	**.10**

Note: Numbers associated with MTD are in bold.

7.4.3 The Case for an Automated Process

Overall, the design experience of the bortezomib trial and t
mix of conventions, intuition, and validation by simulatio
process was guided by objective criteria such as model sensi
the use of these criteria was not systematic. This trial-and
obvious shortcomings. First, the process is time consuming
system is quite complex, proper design tuning requires a hig
with the method. This prevents widespread use of the CRM. Sec
the final design is quite arbitrary. The third problem, which
difficulty, is that the process is virtually irreproducible
my best recount of what happened at the planning stage of the t
likely to obtain a different CRM design should we go through t
process again. This is highly undesirable and confusing to o
imagine 10 statisticians giving 10 different answers when a
calculation for a single-armed study!

As seen in Shen and O'Quigley [94] and Cheung and Chappell [21
calibrated CRM can lead to poor operating characteristics.
to choose the CRM design parameter carefully. This, togethe
with the above trial-and-error approach, calls for an autom
The flowchart in Figure 7.1 outlines one such process. To illu
flowchart, suppose the clinical investigators have chosen s

and all but the lowest dose are higher than the standard use of no external evidence a priori to suggest that the MTD is among may be prudent to specify $\nu_0 = 1$ in the CRM model and to start the trial at $x_1 =$ According to the flowchart in Figure 7.1, we may consult the te for model calibration.

The scheme in Figure 7.1 is not to be used as a substitute for co between the study statistician and the clinical investigat such communication. In this sense, the flowchart presents a s instead of complete automation. For example, the choice of the sample size N be revisited after all other design parameters are specified is to calibrate a CRM design over a range of feasible sample si offs between accuracy and cost can be presented to the clinic final decision.

Also, the presented approach is a pragmatic choice and is by n nor the only logical automated approach. While the flowchart start menu" for statisticians new to the CRM, it may also serv for an informed trial-and-error approach done by statistic CRM.

Chapter 8

8.1 Introduction

This chapter discusses the calibration of the initial guess
associated with the test doses, or the skeleton. Section 8.2
which the initial guesses are specified via a single design pa
width δ of the indifference intervals of the CRM model. Section 8.3 d
can be chosen for a CRM application, and gives some numerical
for the logistic model. Section 8.4 applies the techniques t
trial as a case study.

8.2 Half-width (δ) of Indifferent Interval

As mentioned in the earlier chapters, the dose label d_k
via backward substitution by solving $p_{0k} = F(d_k, \hat{\beta}_0)$ where $\hat{\beta}_0$
the model parameter β and p_{0k}
ciated with dose level k. Ideally, the parameter p_{0k}
prior belief about dose level k
the clinical investigators, except perhaps at the prior MTD d_{ν_0}
$p_{0\nu_0} = \theta$ so that

$$\theta = F(d_{\nu_0}, \hat{\beta}_0).$$

Even if the clinical investigators attempt to provide their
the resulted CRM model may fail to meet the regularity condit
see Exercise 8.1. Therefore, it is pragmatic to treat each p_{0k}
calibrated so that the CRM will yield good operating charact
p_{0k} for a dose other than ν_0
on the sensitivity of a CRM model. Recall that the indifferen
model under a dose–toxicity configuration with MTD $\nu \in \{2, \ldots, K-1\}$
$(F(d_{\nu-1}, b_\nu), F(d_{\nu+1}, b_{\nu+1}))$, where b_j
Chapter 5. Let

$$\delta_\nu = \frac{F(d_{\nu+1}, b_{\nu+1}) - F(d_{\nu-1}, b_\nu)}{2}$$

denote the half-width of the interval. For a given ν, a CRM design $\mathscr{D}$ will converge to a dose close to the true MTD ν

$$P_{\mathscr{D}}\{x_n \in \mathscr{I}_\pi(p-\delta_\nu, p+\delta_\nu) \text{ for all } n \geq N\} = 1$$

for some $N > 0$; see displayed equation (5.3). Suppose we can prespecify a half-width δ that we would like to achieve for all ν, that is, $\delta_\nu \equiv \delta$ ference intervals are invariably $\theta \pm \delta$. Then we can compute $\{d_1, \ldots, d_K\}$ by solving

$$F(d_{\nu-1}, b_\nu) = \theta - \delta \text{ and } F(d_{\nu+1}, b_{\nu+1}) = \theta + \delta$$

for $\nu = 2, \ldots, K-1$, along with b_νs in accordance with Definition (5.5):

$$F(d_{\nu-1}, b_\nu) + F(d_\nu, b_\nu) = 2\theta.$$

Simplifying (8.2) and (8.3) gives the following algorithm:

Algorithm 8.1. Initial guesses of toxicity probabilities via δ

1. Calculate d_{ν_0} by solving (8.1).
2. If $\nu_0 = 1$, skip this step; otherwise, iterate ν from ν_0 down to 2:
   (a) Calculate b_ν by solving $F(d_\nu, b_\nu) = \theta + \delta$;
   (b) Calculate $d_{\nu-1}$ by solving $F(d_{\nu-1}, b_\nu) = \theta - \delta$.
3. If $\nu_0 = K$, skip this step; otherwise, iterate ν from $\nu_0 + 1$ up to K:
   (a) Calculate b_ν by solving $F(d_{\nu-1}, b_\nu) = \theta - \delta$;
   (b) Calculate d_ν by solving $F(d_\nu, b_\nu) = \theta + \delta$.
4. Set $p_{0k} = F(d_k, \hat{\beta}_0)$ for $k = 1, \ldots, K$.

Consider the logistic model (3.3) with $a_0 = 3$ and a prior mean 0 for β $\hat{\beta}_0 = 0$, for a trial with $\theta = 0.25, K = 5$, and $\nu_0 = 3$. If we set $\delta = 0.$ of Algorithm 8.1 yields $d_3 = -4.10$ by solving (8.1):

$$0.25 = \frac{\exp\{3 + \exp(0) d_3\}}{1 + \exp\{3 + \exp(0) d_3\}}.$$

Iterating ν in Step 2 of the algorithm further gives, for $\nu = \nu_0 = 3$,

$$b_3 = \log\left[\{\text{logit}(0.25 + 0.07) - 3\}/d_3\right] = -0.088$$

and

$$d_2 = \{\text{logit}(0.25 - 0.07) - 3\}/\exp(b_3) = -4.93;$$

and, $b_2 = -0.273$ and $d_1 = -5.93$ for $\nu = 2$.

Likewise, we obtain $b_4 = 0.097, b_5 = 0.282$, $d_4 = -3.41$, and $d_5 = -2.$ iterating $\nu_0 = 4, 5$ in Step 3. Finally, applying Step 4 gives $p_{01} = 0.05$, $p_{02} = 0$ $p_{03} = 0.25, p_{04} = 0.40$, and $p_{05} = 0.54$.

Two technical points are noteworthy. First, since the indif
asymptotic criterion, the prior distribution does not play
δ. Second, Algorithm 8.1 is invariant to the parameterizatio
model, that is, the function $c(\beta)$
all ψ-equivalent models.

Algorithm 8.1 is implemented for the empiric (3.2) and the lo
by the R function `getprior`
CRM model can be obtained by the following R codes:

```
> # Get initial guesses with half-width del = 0.07
> library(dfcrm)
> del <- 0.07
> theta <- 0.25
> nu0 <- 3
> p0 <- getprior(del,theta,nu0,K,model="logistic")
> round(p0,digits=2)
[1] 0.05 0.13 0.25 0.40 0.54
>
```

Choosing the initial guesses based on a prespecified sensitivity δ
oretical basis for the calibration process. At the same time
calibration process much more manageable by reducing the nu
eters from K (p_{0k}s) to 1 (δ
the choice of the prior MTD ν_0
rameter or in accordance with (7.2) in the previous chapter.
question is how δ should be chosen.

8.3 Calibration of δ

8.3.1 *Effects of δ on the Accuracy Index*

A δ-sensitive CRM will eventually select a dose with toxicity p
between $\theta \pm \delta$. Therefore, a small value δ
On the other hand, in the finite sample setting, the CRM with a smaller δ
necessarily perform better, especially when the true MTD is close to the prior ν_0
B_j^δ denote the home set for dose j
given δ—the superscript δ in B_j^δ
half-width δ.

Theorem 8.1. *Assume that $F_k(\beta)$ is decreasing in β for all k. Then for $\delta_1 < \delta_2$
$\nu_0 = 2, \ldots, K-1$,*

$$B_1^{\delta_1} \supset B_1^{\delta_2},\ B_{\nu_0}^{\delta_1} \subset B_{\nu_0}^{\delta_2} \text{ and } B_K^{\delta_1} \supset B_K^{\delta_2}.$$

The theorem states that a small value of δ corresponds to a narrow $B_{\nu_0}^\delta$
the definition (5.4) of home sets, the Bayesian CRM will select dose level j
next patient, if $\hat{\beta}_n \in B_j^\delta$
δ will select ν_0 less often than that with a large δ

is a correct guess of the MTD, it is preferable to use a large value of δ
arguments, we can infer that a small δ
be among the extreme ends of the dose range.

8.3.2 The Calibration Approach

The approach in this section is to calibrate the half-width δ
CRM, for given clinical parameters (θ, K, ν_0, N), model functional form F
distribution of β, so that the CRM design will maintain a reasonably high accur
index. As suggested by Theorem 8.1, the ideal choice of δ
MTD ν, as well as the underlying dose–toxicity curve π
simplify the calibration criteria by adopting the notion of
with respect to a calibration set $\mathscr{P}$
weight vector W_π such that $\sum_{\pi \in \mathscr{P}} W_\pi = 1$.

Therefore, let $\mathscr{A}_N(\delta, \pi)$
a half-width δ under dose–toxicity curve π
here the accuracy index (6.1) with respect to the absolute discrepancy $\rho_k = |p_k -$
The risk-adjusted average of accuracy is defined as

$$\bar{\mathscr{A}}_N(\delta) = \sum_{\pi \in \mathscr{P}} W_\pi \mathscr{A}_N(\delta, \pi).$$

By iterating δ from 0.01 to $\theta - 0.$
half-width δ with the largest value of $\bar{\mathscr{A}}_N(\delta)$

While there are many ways to choose the calibration set, a rea
encompass a wide range of scenarios with minimal number of co
use the plateau set $\mathscr{P}^* = \{\pi_l^* : l = 1, \ldots, K\}$ where each $\pi_l^* = (p_1^*, \ldots, p_K^*)^T$

$$p_l^* = \theta, \; \frac{p_l^*/(1-p_l^*)}{p_k^*/(1-p_k^*)} = 2 = \frac{p_{k'}^*/(1-p_{k'}^*)}{p_l^*/(1-p_l^*)} \text{ for } k < l < k',$$

and a uniform weight $W_{\pi_l^*} = 1/K$ for all l
in the plateau set $\mathscr{P}^*$
has an unambiguous MTD ν so that they together account for all possible ν
plateau configurations prescribe difficult scenarios for th
MTD, because the performance of the CRM is expected to improv
dose–toxicity curves. In addition, the toxicity probabili
MTD are specified so that the toxicity odds ratio is 2 when comp
this prescribes modest jump sizes. Table 8.1 lists the plate
scenario with $\theta = 0.25$ and $K =$
below the MTD is 11 percentage points, and above 15 percentag
sizes are modest in light of the typical sample size (20–40) u
and Cheung [60] verify that accuracy will improve if the cali
an odds ratio greater than 2 as in (8.4). The calibration algo
$\mathscr{P}^*$ will thus produce a design with reasonable accuracy under di
so as to give a lower bound of performance for steeper dose–to
CRM model is chosen, we may further evaluate the design's ope
under a set of validation scenarios; see Section 8.4.

Table 8.1 The plateau calibration configurations for $\theta = 0.25$ and $K = 5$

curve	p_1^*	p_2^*	p_3^*	p_4^*	p_5^*	ν
π_1^*	0.25	0.40	0.40	0.40	0.40	1
π_2^*	0.14	0.25	0.40	0.40	0.40	2
π_3^*	0.14	0.14	0.25	0.40	0.40	3
π_4^*	0.14	0.14	0.14	0.25	0.40	4
π_5^*	0.14	0.14	0.14	0.14	0.40	5

8.3.3 *Optimal δ for the Logistic Model*

This section presents the numerical values of the optimal δ
Bayesian CRM using the logistic function (3.3) with $a_0 = 3$ and $\beta \sim N(0,1$
under a variety of commonly used clinical parameters. The ca
respect to the plateau configurations (8.4).

To illustrate how the calibration process operates, the calibration results for
$0.25, K = 5$, and $\nu_0 = 3$ are plotted in Figure 8.1 for $N = 20, 30,$
simulated trials for each N. While the accuracy index depends on δ
manner due to discreteness in the design parameters, some tr

- Accuracy improves with a small value of δ
  ends of the dose range (i.e., MTD = 1 or 5).
- There is a reverse trend when the true MTD is equal to the prior $\nu_0 =$
  is best with large δ.
- Accuracy improves as the sample size N

The first two trends are consistent with what Theorem 8.1 impl
no δ is uniformly best. The last bullet point is in line with asymp
CRM with a large δ
therefore, an increase in sample size may not improve accura
will limit our search of δ by iterating from 0.01 to 0.8θ.

While there is no uniformly best δ
the risk-adjusted average is a unimodal function of δ
accuracy around the maximum point is quite flat. For $N =$
reaches maximum at $\delta = 0.07$ with $\bar{\mathscr{A}}_{20}(0.07) = 0.359$, but the index $\bar{\mathscr{A}}_{20}(\delta)$
within 0.01 of the maximum for δ ranging from 0.05 to 0.11. For $N =$
average accuracy 0.514 is attained when $\delta = 0.05$ with $\delta = 0.04 - 0.$
within 0.01 unit of the index. Therefore, we should view the "optimal" δ
from the calibration algorithm as a typical value among the good δ
a unique optimal choice. More importantly, the CRM model wit
(indicated by a solid dot in Figure 8.1) yields competitive a
individual configurations. In the absence of a uniformly opt
this a reasonable design approach.

It is interesting to note that the range of δ
Figure 8.1 gradually shifts to the left as N

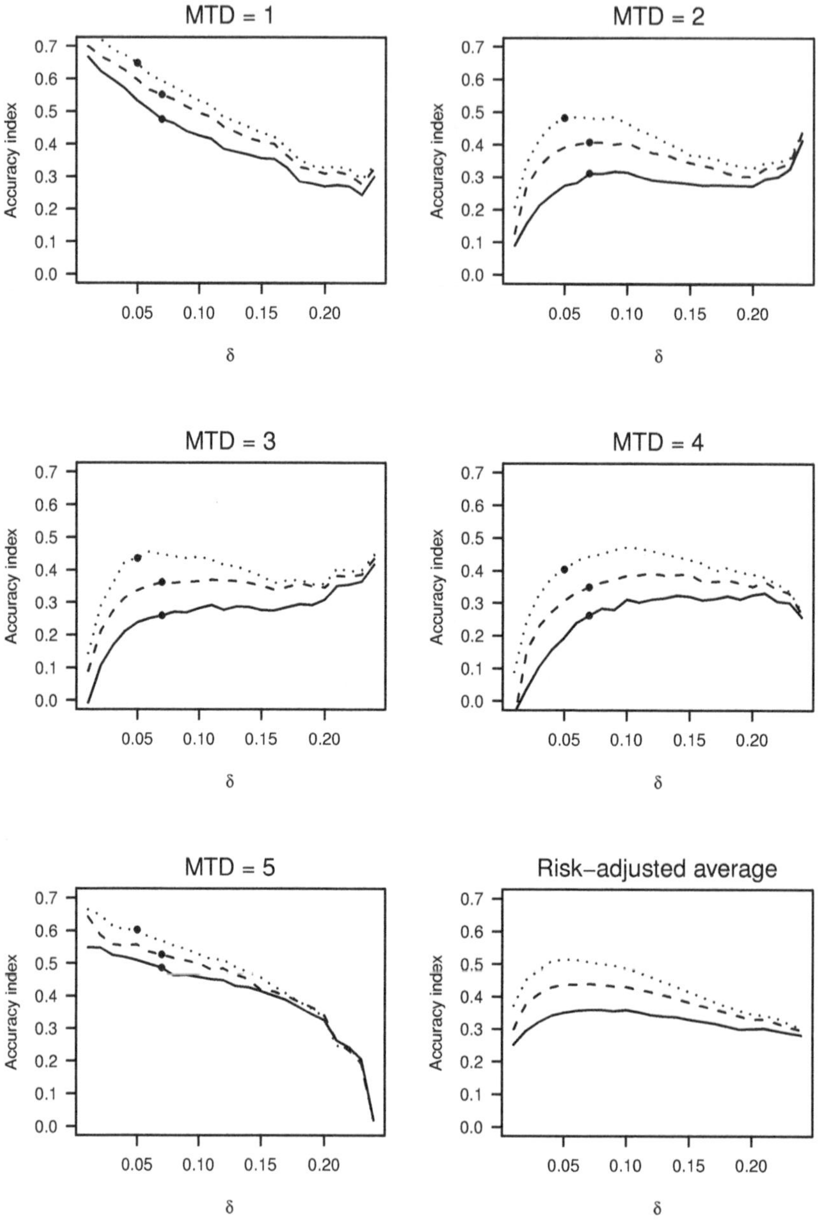

Figure 8.1 Accuracy index versus δ
from 1 to 5, for clinical settings $\theta = 0.25, K = 5, \nu_0 = 3$, and $N =$
and 40 (dotted). The dot on each line indicates the value of δ
average of the accuracy index.

size grows, an asymptotic consideration becomes increasin
small δ (which is asymptotically desirable) is preferred.

We choose the accuracy index (6.1) with the absolute discrep
index of the performance of the CRM and a numerical basis for c
calibration process. However, the PCS is likely a more intui
than an accuracy index. Figure 8.2 plots the PCS under the pla
the risk-adjusted average PCS against δ
in Figure 8.1. In addition, the optimal δ
indicated by solid dots in Figure 8.2, yields competitive op
terms of PCS and the risk-adjusted average of PCS. In fact, th
generally insensitive to the choice of the discrepancy meas
index, when the calibration is done with respect to the plateau calibration set $\mathscr{P}$

Tables 8.2–8.5 show the optimal δ along with the corresponding $\bar{\mathscr{A}}_N(\delta)$
possible combinations of the following clinical parameter
constraint (7.1) is met:

- $\theta = 0.10, 0.20, 0.25, 0.33$
- $K = 4, 5, 6, 7$
- $\nu_0 = 1, 2, \ldots, \lceil K/2 \rceil$
- $N = 20, 25, 30, 35, 40$

These results are obtained on a discrete domain of δ
from 0.01 to 0.8θ. For example, for $\theta = 0.25$, the values of δ
grid $0.01, 0.02, 0.03, \ldots, 0.19, 0.$
function (3.3) with $a_0 = 3$ and $\beta \sim N(0, 1.34)$
empiric model are reported in Lee and Cheung [60], who use the
basis of comparison. Some general trends are observed:

1. The risk-adjusted average accuracy $\bar{\mathscr{A}}_N(\delta)$
   the sample size N, but decreases with the number K of test doses.
2. The range of optimal δ shifts gradually toward small values as N
   does not seem to depend on K.
3. There is no apparent effect of ν_0 on the optimal δ and the accuracy $\bar{\mathscr{A}}_N(\delta)$
   for $\theta = 0.33$ when a higher starting dose ν_0 corresponds to a larger optimal
   higher accuracy $\bar{\mathscr{A}}_N(\delta)$.
4. The optimal δ range is 0.02–0.04 for $\theta = 0.10$; 0.04–0.08 for $\theta = 0.20, 0.$
   0.04–0.10 for $\theta = 0.33$.

8.4 Case Study: The Bortezomib Trial

In this section, we will discuss the calibration of the skele
bortezomib trial. In particular we will consider the empiri
their respective optimal values of δ
presented in Chapter 7, Section 7.4.1.

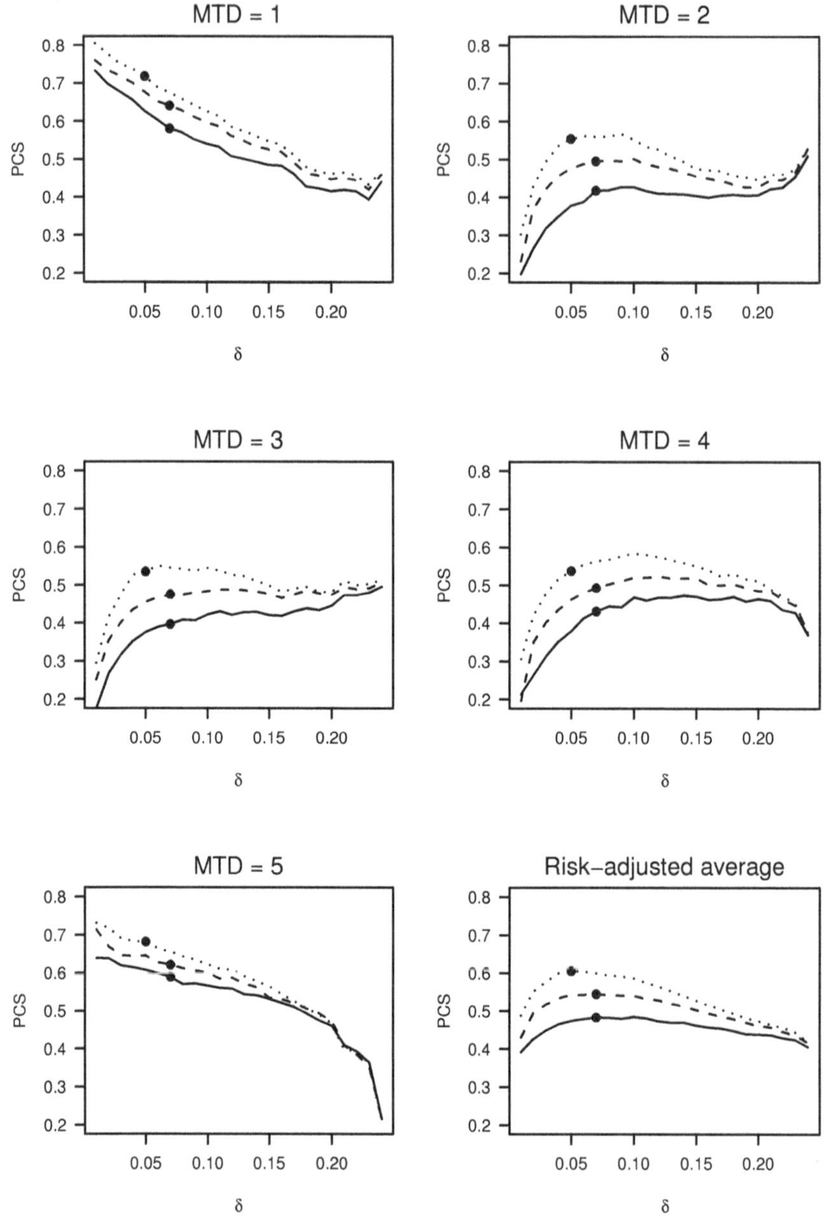

Figure 8.2 PCS versus δ under the plateau configurations (8.4) with MTD ranging from
for clinical settings $\theta = 0.25, K = 5, \nu_0 = 3$, and $N =$
The dot on each line indicates the value of δ
accuracy index.

Table 8.2 Optimal δ and the corresponding $\bar{\mathscr{A}}_N(\delta)$ for the logistic model (3.3) with a and $\beta \sim N(0, 1.34)$ given $\theta = 0.10$, number of dose level (K), prior MTD (ν_0
size (N)

K	ν_0	N 20	25	30	35	40
4	1			(.04, .32)	(.03, .35)	(.03, .38)
	2			(.04, .32)	(.04, .36)	(.03, .38)
5	1			(.03, .29)	(.03, .31)	(.03, .34)
	2			(.03, .29)	(.03, .32)	(.03, .34)
	3			(.03, .28)	(.02, .31)	(.02, .34)
6	1				(.03, .29)	(.03, .31)
	2				(.02, .29)	(.03, .31)
	3				(.02, .29)	(.02, .31)
7	1				(.03, .27)	(.03, .30)
	2				(.03, .27)	(.02, .29)
	3				(.02, .27)	(.02, .29)
	4				(.02, .26)	(.02, .29)
Range of δ				.03–.04	.02–.04	.02–.03

Table 8.3 Optimal δ and the corresponding $\bar{\mathscr{A}}_N(\delta)$ for the logistic model (3.3) with a and $\beta \sim N(0, 1.34)$ given $\theta = 0.20$, number of dose level (K), prior MTD (ν_0
size (N)

K	ν_0	N 20	25	30	35	40
4	1	(.06, .35)	(.07, .39)	(.05, .43)	(.06, .46)	(.05, .50)
	2	(.08, .36)	(.06, .40)	(.06, .44)	(.06, .48)	(.06, .51)
5	1	(.06, .32)	(.07, .36)	(.04, .40)	(.05, .43)	(.04, .46)
	2	(.07, .33)	(.07, .37)	(.06, .41)	(.06, .44)	(.06, .48)
	3	(.07, .33)	(.06, .36)	(.06, .40)	(.05, .44)	(.05, .47)
6	1		(.06, .33)	(.04, .37)	(.05, .41)	(.04, .43)
	2		(.07, .35)	(.06, .38)	(.06, .42)	(.04, .45)
	3		(.05, .35)	(.05, .38)	(.05, .42)	(.05, .45)
7	1		(.07, .32)	(.05, .35)	(.05, .38)	(.04, .42)
	2		(.06, .33)	(.06, .37)	(.06, .40)	(.04, .43)
	3		(.07, .33)	(.06, .37)	(.05, .40)	(.04, .43)
	4		(.07, .32)	(.05, .36)	(.04, .39)	(.05, .42)
Range of δ		.06–.08	.05–.07	.04–.06	.04–.06	.04–.06

Table 8.4 Optimal δ and the corresponding $\bar{\mathscr{A}}_N(\delta)$ for the logistic model (3.3) with a_0 and $\beta \sim N(0, 1.34)$ given $\theta = 0.25$, number of dose level (K), prior MTD (ν_0 size (N)

K	ν_0	N				
		20	25	30	35	40
4	1	(.06, .38)	(.05, .41)	(.06, .46)	(.06, .50)	(.05, .54)
	2	(.07, .39)	(.08, .44)	(.07, .47)	(.07, .51)	(.06, .55)
5	1	(.07, .35)	(.06, .39)	(.06, .43)	(.05, .46)	(.05, .50)
	2	(.06, .36)	(.08, .40)	(.07, .45)	(.07, .48)	(.06, .51)
	3	(.07, .36)	(.08, .42)	(.07, .44)	(.06, .48)	(.05, .51)
6	1	(.07, .32)	(.05, .36)	(.05, .41)	(.05, .44)	(.04, .47)
	2	(.07, .34)	(.06, .38)	(.07, .42)	(.06, .45)	(.06, .49)
	3	(.07, .34)	(.06, .38)	(.06, .42)	(.06, .46)	(.05, .49)
7	1		(.05, .34)	(.05, .39)	(.05, .41)	(.04, .44)
	2		(.06, .36)	(.05, .40)	(.05, .43)	(.06, .46)
	3		(.06, .36)	(.06, .41)	(.06, .44)	(.05, .47)
	4		(.07, .36)	(.07, .40)	(.07, .43)	(.05, .46)
Range of δ		.06–.08	.05–.08	.05–.07	.05–.07	.04–.06

Table 8.5 Optimal δ and the corresponding $\bar{\mathscr{A}}_N(\delta)$ for the logistic model (3.3) with a_0 and $\beta \sim N(0, 1.34)$ given $\theta = 0.33$, number of dose level (K), prior MTD (ν_0 size (N)

K	ν_0	N				
		20	25	30	35	40
4	1	(.07, .40)	(.07, .45)	(.07, .49)	(.07, .53)	(.07, .57)
	2	(.10, .43)	(.09, .47)	(.09, .51)	(.07, .55)	(.06, .58)
5	1	(.07, .37)	(.07, .41)	(.06, .45)	(.05, .49)	(.05, .53)
	2	(.07, .39)	(.07, .43)	(.07, .47)	(.07, .52)	(.06, .55)
	3	(.10, .40)	(.09, .44)	(.09, .48)	(.07, .51)	(.06, .55)
6	1	(.07, .35)	(.07, .38)	(.05, .42)	(.05, .46)	(.04, .50)
	2	(.07, .37)	(.07, .41)	(.06, .44)	(.07, .48)	(.06, .52)
	3	(.09, .38)	(.09, .42)	(.07, .45)	(.06, .49)	(.06, .53)
7	1		(.06, .36)	(.05, .40)	(.05, .43)	(.04, .47)
	2		(.07, .38)	(.06, .42)	(.06, .45)	(.05, .49)
	3		(.09, .40)	(.06, .43)	(.06, .47)	(.06, .50)
	4		(.09, .40)	(.07, .43)	(.07, .47)	(.06, .50)
Range of δ		.07–.10	.06–.09	.05–.09	.05–.07	.04–.07

The optimal empiric model.
for δ similar to that in Section 8.3.2 for the empiric model used in
study, and recommend $\delta = 0.10$. Using the function getprior

```
> p0 <- getprior(0.10, 0.25, 3, 5, model="empiric")
```

we get $p_{01} = 0.01, p_{02} = 0.08, p_{03} = 0.25, p_{04} = 0.46$, and $p_{05} = 0.$
represents a slightly steeper prior than that of the bortezomib study model.

The optimal logistic model. To use the logistic model (3.3) with $a_0 =$
refer to Table 8.4. For $K = 5, \nu_0 = 3$, and $N =$
of δ is 0.07. Using the function getprior

```
> p0 <- getprior(0.07, 0.25, 3, 5, model="logistic")
```

we get $p_{01} = 0.05, p_{02} = 0.12, p_{03} = 0.25, p_{04} = 0.40$, and $p_{05} = 0.$
be a coincidence that this set of initial guesses is almost id
model and that both have similar indifference intervals.

While the calibration of the optimal empiric and logistic mo
respect to the plateau configurations, it is important to val
under realistic dose–toxicity scenarios. These validatio
in collaboration with the clinical investigators. The vali
crucial for the bortezomib trial because the plateau configu
general expectation of dose–toxicity relationship of an an
compares the performance of the study model and the two optim
the dose–toxicity configurations used in the original study
Overall, the three designs have comparable accuracy with ab
under all scenarios: the average PCS over the five scenarios i
respectively, for the study model, the optimal empiric, and
the average accuracy index $\mathscr{A}_{20}$
of the nonparametric optimal design is also provided in the t
nonparametric optimal design is generally superior to the C
PCS 0.76 and average $\mathscr{A}_{20} = 0.$
example, the efficiency of the study model in terms of average
as

$$\frac{0.65}{0.76} \times 100\% \approx 86\%.$$

Similarly, the efficiencies of the three CRM designs are appr
of the nonparametric optimal design.

From a safety viewpoint, the optimal empiric model is prefer
overdose numbers are below that of balanced randomization i
In contrast, the study model and the optimal logistic model t
patients at dose level 5 (an overdose) under Scenario 4.

The variability of the PCS across the scenarios is also notew
dose–toxicity curve is unknown, it will be reassuring to loo
PCS and verify that the method's performance is adequate acr
this regard, the study model is *slightly*
and 13 percentage points, respectively, for the study model
optimal logistic; and, the respective standard deviations

Table 8.6 The distribution of dose selection, average toxic
number (OD) of three CRM models for the bortezomib trial with $\theta = 0.25$, $K = 5$, $\nu_0 =$
$N = 20$

Model	Probability of selecting dose					$\mathscr{A}_{20}$	ATN	OD
Scenario 1:	**.25**	.40	.45	.55	.60			
NP Optimal	**.78**	.16	.05	.01	.00	.82	—	—
The study model	**.67**	.25	.08	.01	.00	.72	6.9	9.5
Optimal empiric	**.58**	.31	.10	.01	.00	.65	7.2	11.3
Optimal logistic	**.69**	.23	.07	.01	.00	.74	6.7	8.5
Scenario 2:	.05	**.25**	.40	.45	.55			
NP Optimal	.07	**.71**	.16	.05	.01	.70	—	—
The study model	.12	**.58**	.25	.05	.00	.57	5.8	8.4
Optimal empiric	.08	**.60**	.26	.06	.00	.60	6.0	8.7
Optimal logistic	.12	**.57**	.24	.06	.01	.56	5.6	8.2
Scenario 3:	.05	.05	**.25**	.45	.55			
NP Optimal	.00	.07	**.78**	.13	.01	.76	—	—
The study model	.00	.15	**.68**	.16	.01	.64	5.2	5.1
Optimal empiric	.00	.12	**.71**	.16	.01	.68	5.2	4.8
Optimal logistic	.00	.16	**.64**	.18	.01	.59	5.1	5.4
Scenario 4:	.05	.05	.08	**.25**	.45			
NP Optimal	.00	.00	.12	**.74**	.14	.68	—	—
The study model	.00	.01	.19	**.64**	.16	.57	4.6	4.2
Optimal empiric	.00	.01	.20	**.67**	.12	.61	4.5	3.1
Optimal logistic	.00	.02	.19	**.61**	.18	.53	4.8	4.8
Scenario 5:	.05	.05	.08	.12	**.25**			
NP Optimal	.00	.00	.01	.19	**.79**	.80	—	—
The study model	.00	.01	.05	.28	**.66**	.67	3.6	0.0
Optimal empiric	.00	.01	.06	.34	**.60**	.60	3.4	0.0
Optimal logistic	.00	.01	.04	.24	**.70**	.71	3.7	0.0

Note: The results for the nonparametric optimal design (NP o
references. Numbers associated with the MTD are in bold.

0.05. It should be noted, however, that comparing variabili
only when the methods are comparable in terms of average PCS.
be explored further in Chapter 9. Scenario-by-scenario, th
having a larger half-width δ
other two when the true MTD is equal to the prior MTD $\nu_0 =$
when the true MTD is on the extreme ends of the dose range. Howe
performance of the three models is very comparable.

Based on the accumulative simulation experience from a numb
the CRM models obtained by the calibration process generall
operating characteristics against the corresponding stud
in light of the convenience the automated calibration appro

calibration process is much less labor-intensive than the t
can be programmed in a few lines of R code with help from the 'df
process is mildly computer-intensive and may take a day or tw
run. For the most common clinical settings, the δ
usually give reasonable designs.

8.5 Exercises and Further Results

Exercise 8.1. For a trial with $K = 5$ dose levels and a target $\theta = 0.$
investigator believes that the highest dose is the MTD a prio
are virtually nontoxic. Thus, he provides $p_{01} = p_{02} = 0.01$, $p_{03} = p_{04} = 0.05$, $p_{05} = 0.10$. Evaluate the dose labels for the five levels if the empiric
with $\hat{\beta}_0 = 0$. Comment on this skeleton.

Exercise 8.2. Prove the following properties about skeletons obtained by Algorithm 8.1 with a half-width δ:

a. The skeleton is strictly increasing, that is, $p_{01} < \cdots < p_{0K}$.

b. $|p_{0k} - \theta|$ is strictly increasing in δ (Lee and Cheung [60], Theorem 1).

The result in part b implies that a large δ
δ to a shallow one. As such, a large δ
input, because the prior MTD ν_0
guess p_{0k} is much different from $p_{0,\nu_0} = \theta$ for $k \neq \nu_0$. Likewise, a small δ
skeleton) indicates little difference between the prior MTD ν_0
terms of initial guesses of toxicity probabilities, and may
input.

Chapter 9

Least Informative Normal Prior

9.1 Introduction

This chapter discusses the specification of the prior standard deviation σ_β
model parameter β
informative prior in the context of dose finding, and summari
the specification of prior distribution in terms of its varia
calibration approaches: Section 9.3 presents an approach i
least informative prior for a given sensitivity δ
informative model with respect to δ
context of the bortezomib trial.

9.2 Least Informative Prior

9.2.1 *Definitions*

As discussed in Chapter 4, a Bayesian CRM using model (4.1) with $c(\beta) =$
is invariant to the mean of a normal prior distribution. Thus
zero normal prior

$$\beta \sim N(0, \sigma_\beta^2)$$

without loss of generality. As a result, the prior distribut
by the standard deviation σ_β
σ_β corresponds to a vague prior, whereas a small σ_β
the context of dose finding, however, the notion of a noninfor
defined with reference to our belief about the MTD, rather tha
parameter β.

Let μ_j denote the probability that dose j
distribution, that is, $\mu_j \equiv \Pr(\beta \in B_j) = \Pr(\nu_\beta = j)$, where ν_β
MTD when β is the true parameter value, so that $\nu_\beta = j$ if and only if β
to the home set B_j. Moreover, since the home sets are disjoint intervals define
(5.5), the prior mass $\mu_j = G(b_{j+1}) - G(b_j)$ for $j = 1, \ldots, K$, where $b_1 \equiv -\infty$
$b_{K+1} \equiv \infty$. For the mean zero normal prior (9.1), we have

$$\mu_j = \Phi\left(b_{j+1}/\sigma_\beta\right) - \Phi\left(b_j/\sigma_\beta\right)$$

for $j = 1, \ldots, K$, where Φ

In addition, the first two central moments of ν_β are

$$E(\nu_\beta) \equiv e_\nu = \sum_{j=1}^{K} j\mu_j \text{ and } \mathrm{var}(\nu_\beta) \equiv \sigma_\nu^2 = \sum_{j=1}^{K} j^2\mu_j - e_\nu^2.$$

Table 9.1 shows the distribution of ν_β for the logistic model (3.3) with a_0
$\delta = 0.07$, and various values of σ_β for a trial with $\theta = 0.25$, $K = 5$, and $\nu_0 =$
distribution of ν_β is unimodal with peak at $\nu_0 = 3$ when $\sigma_\beta = 0.$
turns U-shaped as σ_β increases. For example, when $\sigma_\beta = 1.$
used in the optimal logistic model in Chapter 8, Section 8.4)
either dose 1 or dose 5 is the MTD with over 80% probability. Th
a large value of σ_β is by no means vague. In contrast, when $\sigma_\beta = 0.$
distribution of ν_β is quite flat, and represents ignorance about the true MTD a pr
Thus, a noninformative prior in terms of ν_β
that is, $\mu_j = 1/K$ for all j
close μ_js is to a uniform distribution.

Table 9.1 The distribution of ν_β for the logistic model (3.3) with $a_0 = 3$ and $\delta = 0.$
trial with target $\theta = 0.25$, $K = 5$, and $\nu_0 = 3$

j	(b_j, b_{j+1})	μ_j with $\sigma_\beta =$			
		0.20	0.33	0.50	1.16
1	$(-\infty, -0.27)$	0.09	0.21	0.29	0.41
2	$(-0.27, -0.09)$	0.24	0.19	0.14	0.06
3	$(-0.09, 0.10)$	0.36	0.22	0.15	0.06
4	$(0.10, 0.28)$	0.23	0.19	0.14	0.06
5	$(0.28, \infty)$	0.08	0.20	0.29	0.40
e_ν		2.98	2.98	2.99	2.99
σ_ν		1.07	1.41	1.61	1.84

Among the many ways to measure the distance between two distr
focus on an approach that matches the standard deviations of two distributions.

Definition 9.1 (least informative normal prior). *For any given CRM model $F_k($ the normal prior (9.1) is said to be least informative if the corresponding σ_ν defined in (9.2) equals the standard deviation of a uniform distribution on $\{1,\ldots,$ that is, $\sigma_\nu = \sqrt{(K^2-1)/12}.$*
β is denoted by σ_β^{LI}.

For example, for $K =$
to $\sigma_\nu = \sqrt{2} = 1.41$. For the logistic model in Table 9.1, we see that $\sigma_\beta^{LI} = 0.$
the corresponding $\mu_j \approx 0.20$ for all j.

A natural alternative to compare μ_j

Kullback-Leibler divergence criterion:

$$D_{KL}(\mu \mid unif) = \sum_{j=1}^{K} \mu_j \log(K\mu_j).$$

A noninformative normal prior can be obtained by choosing σ_β
"distance" $D_{KL}(\mu \mid unif)$, or equivalently, maximizes the *entropy*
μ_js: $-\sum_{j=1}^{K} \mu_j \log \mu_j$. The value of σ_β
be called the maximum entropy standard deviation. See Berge

Figure 9.1 shows σ_v (left) and D_{KL} (right) against σ_β
all the three models shown in the figure, the least informativ
standard deviations turn out to be identical. In general, th
Definition 9.1 is very similar to that by maximizing entropy.
increases as σ_β increases; thus, the least informative σ_β^{LI}
binary search. For these reasons, we will focus on the least i
now on.

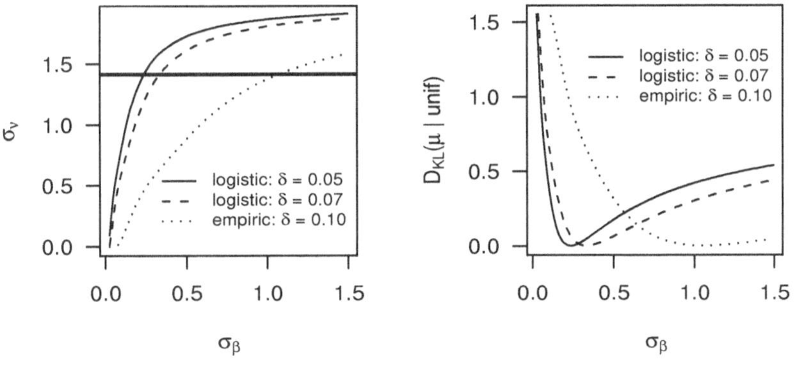

Figure 9.1 Left: σ_v versus σ_β for the logistic model (3.3) with $a_0 = 3$ and $\delta = 0.05, 0.$
the empiric model (3.2) with $\delta = 0.10$ under the setting with $\theta = 0.25, K = 5$, and v_0
The least informative standard deviation is the σ_β value that gives $\sigma_v = 1.$
the dark solid line. Right: The Kullback-Leibler divergence against σ_β
specifications. The maximum entropy standard deviation is the σ_β
Kullback-Leibler divergence criterion (9.3).

9.2.2 *Rules of Thumb*

From Figure 9.1, we observe that the value of least informati
varies with the CRM models. This underlines the fact that the parameter β
different scales under different models. Therefore, a given value of σ_β

different degrees of informativeness in different CRM mode informative standard deviation hence provides some useful of "vagueness":

Definition 9.2. *The normal prior distribution (9.1) is said to be vague for a g CRM model if* $\sigma_\beta \approx \sigma_\beta^{LI}$.

Definition 9.3. *The normal prior distribution (9.1) is said to have large var for a given CRM model if* $\sigma_\beta \gg \sigma_\beta^{LI}$.

Definition 9.4. *The normal prior distribution (9.1) is said to be informativ* ν_0 *for a given CRM model if* $\sigma_\beta \ll \sigma_\beta^{LI}$.

Table 9.2 shows the least informative σ_β^{LI} for the logistic model (3.3) with a_0 for $\theta = 0.25$ under some common design parameters. For this logistic mo least informative σ_β^{LI}
logistic model presented in Section 8.4 is used with $\sigma_\beta = 1$.
prior with large variation.

As a rule of thumb, when the trial has a low starting dose level (e.g., $x_1 =$ and the specified prior MTD ν_0
may be used in order to ensure flexibility for the CRM to reach t
within a reasonable sample size (cf. Table 9.1). On the other
is close to the median dose level (7.2), a vague prior will lik
informative prior favoring ν_0
prior evidence that ν_0

Table 9.2 Least informative standard deviation σ_β^{LI} for the logistic model (3.3) with a_0 for given $\theta = 0.25$, number of dose level (K), prior MTD (ν_0), and half-width (δ indifference intervals

K	$\sigma_{uniform}$	ν_0	δ				
			0.04	0.05	0.06	0.07	0.08
4	1.118	1	0.248	0.312	0.376	0.441	0.507
		2	0.169	0.212	0.255	0.299	0.344
5	1.414	1	0.301	0.378	0.456	0.534	0.614
		2	0.223	0.279	0.337	0.395	0.454
		3	0.189	0.237	0.285	0.334	0.383
6	1.708	1	0.355	0.446	0.537	0.630	0.724
		2	0.278	0.349	0.420	0.492	0.566
		3	0.229	0.288	0.346	0.406	0.466
7	2.000	1	0.410	0.514	0.620	0.727	0.835
		2	0.333	0.418	0.504	0.590	0.679
		3	0.277	0.347	0.418	0.491	0.564
		4	0.256	0.321	0.387	0.453	0.521

9.3 Calibration of σ_β

9.3.1 Calibration Criteria

One may adopt a similar calibration approach to that in Chapt
all other design parameters $(\theta, K, \nu_0, N, F, p_{0k}\text{s})$, and iterate σ_β
as to obtain a value that yields high accuracy. More precisely, let $\mathscr{A}_N(\sigma_\beta, \pi)$
the accuracy index of a one-stage Bayesian CRM assuming σ_β
curve π and define the risk-adjusted average accuracy as

$$\bar{\mathscr{A}_N}(\sigma_\beta) = \sum_{\pi \in \mathscr{P}^*} W_\pi \mathscr{A}_N(\sigma_\beta, \pi)$$

where $\mathscr{P}^*$ is the plateau configurations (8.4) and $W_\pi \equiv 1/K$.

The left panel of Figure 9.2 plots the risk-adjusted average
σ_β for the logistic model (3.3) with $a_0 =$
setting $\theta = 0.25, K = 5, \nu_0 = 1, 2, 3$, and $N =$
σ_β from 0.10 to 10.00 on a discrete grid with width 0.05, but the p
a partial range with prominent features so as to highlight th
The average accuracy index becomes quite flat and remains clo
once σ_β is large enough. For example, for the logistic model with $\delta = 0.$
$\nu_0 = 3$, the average accuracy ranges from 0.35 to 0.39 for $\sigma_\beta \geq 0.$
the risk-adjusted average criterion (9.4) does not appear t
for choosing σ_β.

The right panel of Figure 9.2 plots the standard deviation $\mathscr{S}_N(\sigma_\beta)$
index against σ_β, where

$$\mathscr{S}_N(\sigma_\beta) = \sum_{\pi \in \mathscr{P}^*} W_\pi \left\{\mathscr{A}_N(\sigma_\beta, \pi) - \bar{\mathscr{A}_N}(\sigma_\beta)\right\}^2.$$

It is noteworthy that the variability $\mathscr{S}_N(\sigma_\beta)$
its dependence on σ_β when compared to $\bar{\mathscr{A}_N}(\sigma_\beta)$
maximum, a small variability is desirable because it implie
uniformly across dose–toxicity scenarios with the true MTD
Hence, an alternative calibration criterion is to choose σ_β that minimizes $\mathscr{S}_N($
The criterion based on (9.5) appears to be more sensitive to σ_β
that based on (9.4). For the logistic model with $\delta = 0.07$ and $\nu_0 =$
the standard deviation $\mathscr{S}_N(\sigma_\beta)$
$0.3 \leq \sigma_\beta \leq 1.0$.

9.3.2 An Application to the Choice of ν_0

Figure 9.2 suggests that setting ν_0
the empiric model, the standard deviation $\mathscr{S}_N(\sigma_\beta) \geq 0.15$ for all σ_β with ν_0
whereas $\mathscr{S}_N(\sigma_\beta)$ is uniformly below 0.10 for sufficiently large σ_β with $\nu_0 =$
light of the fact that the average accuracies are comparable for different ν_0
to be advantageous to specify the median dose as the prior MTD

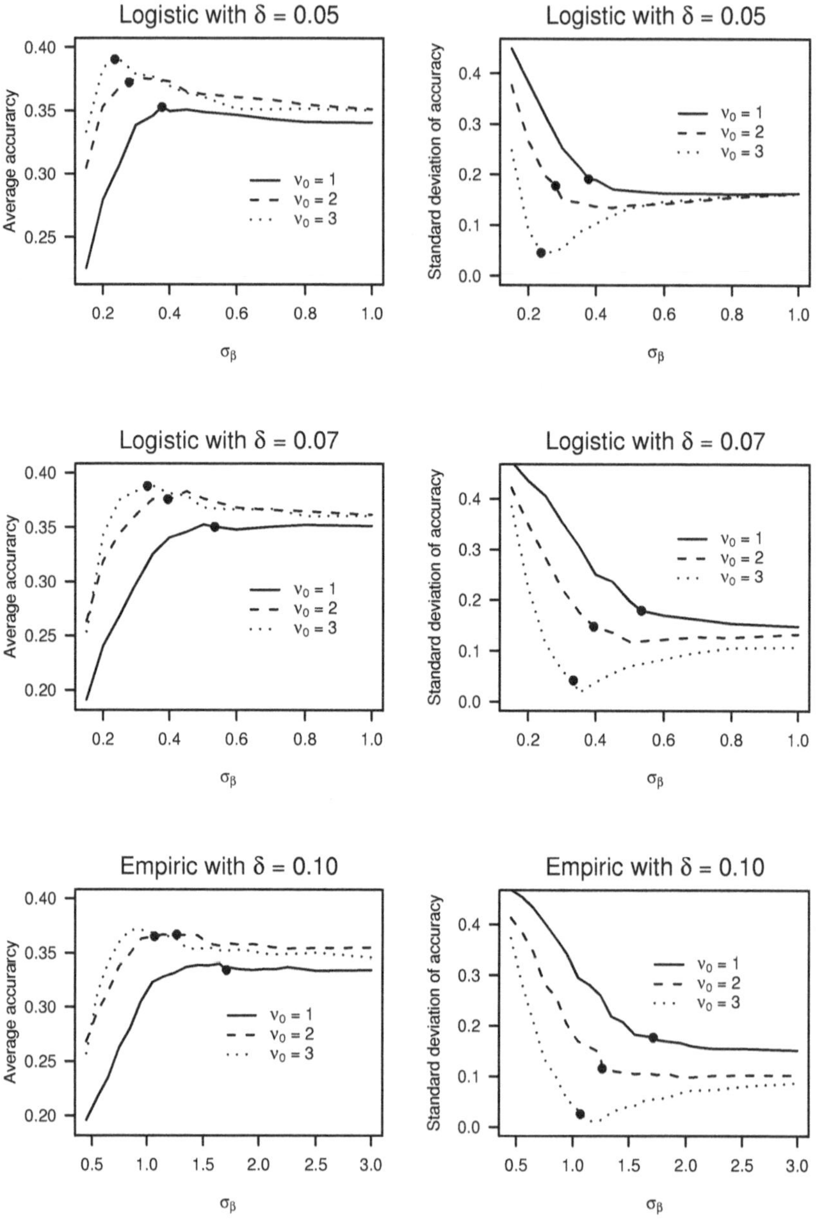

Figure 9.2 Left: Average accuracy $\bar{\mathscr{A}}_N(\sigma_\beta)$ versus σ_β
Table 8.1 for $\theta = 0.25, K = 5, \nu_0 = 1,2,3$, and $N = 20$. Right: $\mathscr{S}_N(\sigma_\beta)$ versus σ_β
dot on each line indicates the value at the least informative σ_β^{LI}.

This is in accordance with the original recommendation by O'
summary, a pragmatic approach is to take ν_0
or equal to the median dose level according to (7.2); this app
view that ν_0 is a model parameter (that is to be tuned) rather than a clinic
(that is to be set with respect to the clinical opinions and constraints).

9.3.3 *Optimality Near* σ_β^{LI}

Interestingly, Figure 9.2 shows that the CRM with a *vague*
in terms of both the average (9.4) and the variability (9.5) criteria.

To offer an explanation, the left panel of Figure 9.3 plots th
index $\mathscr{A}_N(\sigma_\beta, \pi_k^*)$ of the CRM using the logistic model with $\delta = 0.07$ and ν_0
under the plateau configurations (Table 8.1). When σ_β is small, the prior of
unimodal at $\nu_0 = 3$ (cf. Table 9.1), and, therefore, favors dose level 3 a prior
tends to choose dose 3 a posteriori. Thus, the CRM performs well under π_3^*
true MTD is dose 3. For the same reason, however, the CRM shows
to choose the extreme doses under π_1^* and π_5^*
and increasing the variability of the accuracy index. On the
with large variation, the CRM does better under π_1^* and π_5^* than under π_3^*
is around 0.3–0.4, the prior of ν_β
a result, the CRM performs uniformly across all scenarios an
variability $\mathscr{S}_N(\sigma_\beta)$ of the accuracy index.

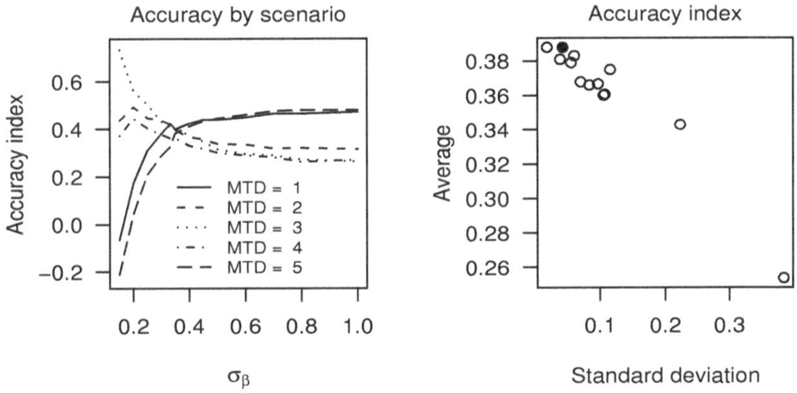

Figure 9.3 Left: $\mathscr{A}_N(\sigma_\nu, \pi_k^*)$ versus σ_β
logistic model with $a_0 = 3$ and $\delta = 0.07$ under $\theta = 0.25, K = 5, \nu_0 = 3$, and $N =$
$\bar{A}_N(\sigma_\nu)$ versus $\mathscr{S}_N(\sigma_\nu)$; the solid dot indicates the least informative prior.

The above reasoning suggests that a *locally* optimal value of σ_β
priors may be a reasonable choice. This approach solves the d
of σ_β in theory increases without bound and that the range of good σ_β

and small $\mathscr{S}_N$) varies with the other design parameters such as the dose–to function. The right panel of Figure 9.3 shows a strong negati
$\mathscr{S}_N(\sigma_\beta)$ and $\bar{\mathscr{A}}_N(\sigma_\beta)$
variability (9.5) are concordant objectives. Therefore, i
mum of σ_β, we will consider both criteria, although the variability i
more discriminatory than the average criterion.

Algorithm 9.1. Locally optimal σ_β near the least informative prior

1. For given clinical parameters (θ, K, ν_0, N) and CRM model $F_k(\beta)$
   least informative σ_β^{LI}.
2. Iterate σ_β in the neighborhood of σ_β^{LI} from $\lambda_1 \sigma_\beta^{LI}$ to $\lambda_2 \sigma_\beta^{LI}$
   having grid width 0.05 for some $\lambda_1 < 1 < \lambda_2$. We suggest $\lambda_1 = 0.8$ and λ_2
3. Choose the standard deviation σ_β
   minimizes the variability of accuracy (9.5) with respect to
   configurations (8.4).

Based on the observed trends about σ_β^{LI}
defined in Algorithm 9.1, we expect the range of locally optimal σ_β
right as

- The number K of test doses increases, or
- The prior MTD ν_0 decreases, or
- The half-width δ of the model's indifference interval increases.

Intuitively, as the number of test doses increases, large va
to allow sufficient flexibility so that the experimentation c
doses in response to the data during the trial. This intuitio
the prior MTD ν_0 being among an extreme end. Finally, a large value of σ_β
to counterbalance the strong prior input due to a large half-width δ
Exercise 8.2).

Example 9.1. Consider the logistic model with $a_0 = 3$, $\delta = 0.07$, $K = 5$, and ν_0
From Table 9.2, we obtain $\sigma_\beta^{LI} = 0.334$. Then applying Algorithm 9.1 with $\lambda_1 \approx$
and $\lambda_2 \approx 1.5$, we iterate σ_β on the grid $\{0.25, 0.30, 0.35, 0, 40, 0.45, 0.50\}$
0.35 as the locally optimal standard deviation based on both cr

Example 9.2. Consider the empiric model with $\delta = 0.10$, $K = 5$, and $\nu_0 =$
least informative $\sigma_\beta^{LI} = 1$.
with width 0.05, the average accuracy (9.4) is maximized when $\sigma_\beta = 0.$
variability (9.5) is minimized when $\sigma_\beta = 1.$
empiric model in Section 8.4 that assumes a normal prior with $\sigma_\beta^2 = 1.34$.

In both examples, the average and variability criteria yiel

choices that are close to the least informative σ_β^{LI}
Algorithm 9.1 reaches the same local optimum regardless of the criterion used.

9.4 Optimal Least Informative Model

It is in principle straightforward to extend the calibratio
the standard deviation σ_β and the sensitivity δ
on a discrete domain over a two-dimensional grid. The comput
however, impose an impractical time constraint.

As illustrated in Example 9.1 and 9.2, the least informative
to the (locally) optimal standard deviation. Therefore, it
the δ that gives the best least informative CRM model:

Algorithm 9.2. Optimal δ among the least informative CRM models

1. For given clinical parameters (θ, K, ν_0, N) and a dose–toxicity function F
   δ from 0.01 to 0.8θ on a discrete domain having grid width 0.01.
2. For each δ and hence a completely specified CRM model $F_k(\beta)$
   informative standard deviation, $\sigma_\beta^{LI}(\delta)$.
3. Choose the pair $(\delta, \sigma_\beta^{LI}(\delta))$ that maximizes $\bar{\mathscr{A}}_N(\sigma_\beta^{LI}(\delta))$
   respect to the calibration configurations (8.4).

The calibration results using Algorithm 9.2 are plotted in F
empiric model (3.2) and the logistic model (3.3) with $a_0 =$
$\theta = 0.25, K = 5, \nu_0 = 3$, and $N = 20$. The dependence of the accuracy index on δ
panel of the figure) demonstrates a different pattern from th
a fixed and large standard deviation is used. For example, under π_5^*
MTD is dose 5, the performance of CRM improves as δ
worsens in Figure 8.1. This is not unexpected. Because the le
deviation is an increasing function of δ
is small, a prior concentrated around $\nu_0 =$
testing and choosing an extreme dose. In contrast, the accur
standard deviation (middle panel of Figure 9.4) in the same m

The lower right corner of Figure 9.4 summarizes the calibrat
the optimal least informative model is specified by $\delta = .06$ with $\sigma_\beta^{LI} = 0.$
logistic, and $\delta = .04$ with $\sigma_\beta^{LI} = 0.$
average accuracy $\bar{\mathscr{A}}_N = 0.$
is not important as long as the other model parameters, namely δ and σ_β
properly. In addition, the average criterion (9.4) seems to
range of good δ from the bad values: the average accuracies of the least info
models are quite flat when δ
for the two models. Therefore, the optimal δ
viewed as a typical value among the good δs.

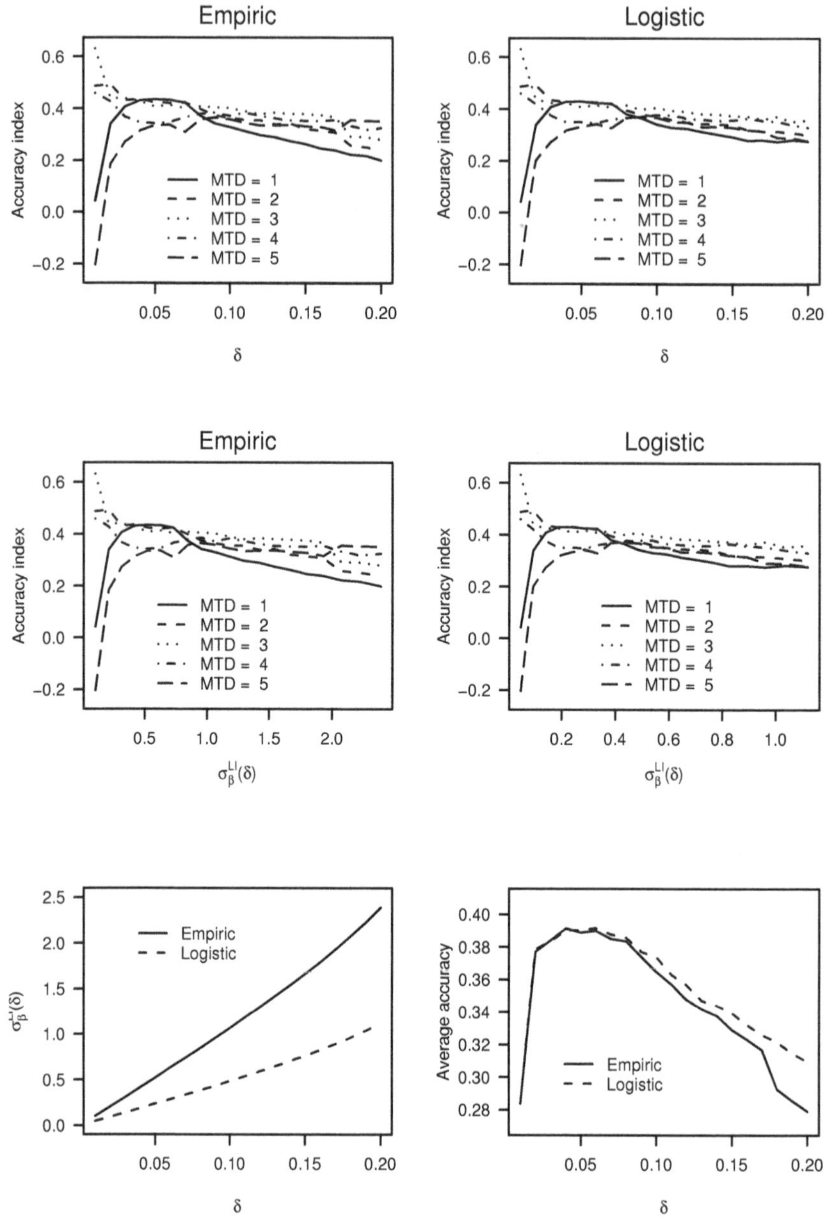

Figure 9.4 Top: Accuracy index versus δ
for the corresponding least informative empiric and logist
$\theta = 0.25, K = 5, \nu_0 = 3$, and $N = 20$. Middle: Accuracy index versus $\sigma_\beta^{LI}(\delta)$
$\sigma_\beta^{LI}(\delta)$ versus δ. Lower right: Average accuracy index versus δ.

Table 9.3 shows the optimal least informative specification $(\delta, \sigma_\beta^{LI}(\delta))$
logistic model (3.3) with $a_0 =$
common clinical parameters. The dependence of the optimal least informative
the sample size N and the number K
Chapter 8 where the standard deviation σ_β
good δ is quite similar to, albeit slightly narrower than, those gi
To summarize:

- For $\theta = 0.10$, the optimal least informative δ range is 0.02–0.03.
- For $\theta = 0.20$, the optimal least informative δ range is 0.03–0.06.
- For $\theta = 0.25, 0.33$, the optimal least informative δ range is 0.04–0.06.

Table 9.3 Optimal least informative specification $(\delta, \sigma_\beta^{LI}(\delta))$
$a_0 = 3$ and the normal prior (9.1) for given θ, K, N

θ	K	N				
		20	25	30	35	40
0.10	4			(.03,.21)	(.02,.14)	(.03,.21)
	5			(.02,.16)	(.02,.16)	(.02,.16)
	6				(.02,.19)	(.02,.19)
	7				(.02,.21)	(.02,.21)
0.20	4	(.05,.23)	(.05,.23)	(.04,.19)	(.06,.28)	(.05,.23)
	5	(.05,.26)	(.04,.21)	(.03,.16)	(.04,.21)	(.05,.26)
	6		(.06,.38)	(.04,.25)	(.05,.32)	(.05,.32)
	7		(.04,.28)	(.04,.28)	(.04,.28)	(.04,.28)
0.25	4	(.04,.17)	(.05,.21)	(.05,.21)	(.06,.26)	(.06,.26)
	5	(.06,.29)	(.06,.29)	(.06,.29)	(.04,.19)	(.04,.19)
	6	(.06,.35)	(.05,.29)	(.05,.29)	(.05,.29)	(.05,.29)
	7		(.05,.32)	(.04,.26)	(.05,.32)	(.04,.26)
0.33	4	(.09,.36)	(.06,.24)	(.06,.24)	(.06,.24)	(.06,.24)
	5	(.06,.27)	(.05,.22)	(.04,.18)	(.06,.27)	(.05,.22)
	6	(.06,.32)	(.06,.32)	(.06,.32)	(.06,.32)	(.06,.32)
	7		(.05,.30)	(.05,.30)	(.05,.30)	(.05,.30)

Note: The prior MTD ν_0 is assumed to be the median dose level, that is, $\nu_0 = 2, 3$ respectively for $K = 4, 5, 6, 7$.

9.5 Revisiting the Bortezomib Trial

This section compares the performance of various models by A
under the validation scenarios in Table 8.6 for the bortezom
includes the following models:

- The study model, the optimal empiric and logistic models in S
  respectively, as models (S), (E0), and (L0).

- The logistic model with locally optimal $\sigma_\beta = 0.35$ for $\delta = 0$. denoted as model (L1).
- The empiric model with locally optimal $\sigma_\beta = 0.95$ for $\delta = 0$. the average accuracy (9.4) criterion in Example 9.2, denote locally optimal σ_β with respect to the variability (9.5) criterion is not consi because it is very similar to model (E0).
- The optimal least informative empiric and logistic models i as models (E2) and (L2) respectively.

The numerical results are shown in Table 9.4, which also give nonparametric optimal design as references. The model (E1) in PCS; there is no reason to expect otherwise since the calib average accuracy (9.4). Using this model, the CRM does very w is dose 3—with a 0.75 PCS when compared to a 0.78 PCS for the non optimal design. However, this design is correct only 54% of t is dose 1. This variation in accuracy can be quite unsettling control nor knowledge about the true MTD. By contrast, model (L0) in terms of variability; this may be due to the fact that (the variability criterion (9.5). In addition, this reducti sacrificing the average accuracy. The optimal least informa yield lower average PCS than that of the study model (E0), but Also, both models bring down the variability in PCS to a compa the nonparametric optimal design. The advantage is clear es the individual PCS under all five scenarios are uniformly abo some assurance of accuracy.

Table 9.4 Operating characteristics of the nonparametric o various model specifications for the bortezomib trial with $\theta = 0.25, K = 5, \nu_0 = 3$, and N under the validation configurations in Table 8.6

Model	δ	σ_β	PCS under Scenario					Ave	Range[a]
			1	2	3	4	5		
(NP)	—	—	.78	.71	.78	.74	.79	.76	.08
(S)	—	1.16	.67	.58	.68	.64	.66	.65	.10
(E0)	0.10	1.16	.58	.60	.71	.67	.60	.63	.13
(E1)	0.10	0.95	.54	.61	.75	.69	. 57	.63	.21
(E2)	0.04	0.42	.61	.61	.68	.63	.58	.62	.10
(L0)	0.07	1.16	.69	.57	.64	.61	.70	.64	.13
(L1)	0.07	0.35	.62	.60	.69	.66	.61	.64	.09
(L2)	0.06	0.29	.61	.61	.69	.65	.61	.63	.08

[a]*Range = maximum PCS − minimum PCS.*

In this simulation, the best performing models are (L1) and (high variation of model (E1) in Table 9.4, we may use the varia with Algorithm 9.1. Finally, the optimal least informative

in the accuracy from the carefully calibrated study model (E
comparable PCS. Thus, the convenience and reproducibility
prove a reasonable design approach; and the optimal least in
for the logistic model is readily available in Table 9.3 for m

Chapter 10

Initial Design

10.1 Introduction

In situations where the clinical investigators believe the
doses but are not comfortable treating patients at a high dos
its lower doses, one may adopt a two-stage CRM that starts a tr
initial sequence $\{x_{i,0}\}$ with $x_{1,0} = d_1$
the choice of the initial sequence. To be specific, Section 10
concepts about the ordering of dose sequences. Section 10.3
approach for $\{x_{i,0}\}$
cal scenarios. Practical issues such as sample size conside
a two-stage CRM will be addressed in Section 10.4. The calibr
be illustrated in Section 10.5 in the context of the NeuSTART.

10.2 Ordering of Dose Sequences

A two-stage CRM assigns doses in accordance with a predeterm
dose sequence $\{x_{i,0}\}$ until the first toxicity is seen, that is, $x_i = x_{i,0}$ for $i \leq$
$Y_j = 1\}$ where

$$x_{i,0} \leq x_{i+1,0}.$$

The primary appeal of a two-stage design is that it allows a lo
is often the lowest dose level, that is, $x_{1,0} = d_1$
sequence should avoid treating many patients at low and inef
which the 3+3 algorithm is criticized. A reasonable strateg
initially and slow down as the trial moves to higher doses [98
implies that the initial cohort sizes are monotone, that is,

$$m_{01} \leq m_{02} \leq \cdots \leq m_{0,K}$$

where

$$m_{0k} = \sum_{i=1}^{N} I(x_{i,0} = d_k)$$

is the *initial cohort size* at dose d_k according to the sequence $\{x_{i,0}\}$
the initial cohort sizes m_{0k}
focus on initial dose sequences that satisfy (10.1) and (10.2).

A fast dose escalation scheme is often perceived to be aggres
conservative. Thus, a notion of relative dose escalation speed is in order:

Definition 10.1 (partial ordering). *Let* $\mathbf{x} = \{x_i\}$ *and* $\tilde{\mathbf{x}} = \{\tilde{x}_i\}$
of doses. Dose escalation in the sequence $\mathbf{x}$
in the sequence $\tilde{\mathbf{x}}$*, denoted by* $\mathbf{x} > \tilde{\mathbf{x}}$ *if* $x_i \geq \tilde{x}_i$
for some i.

To illustrate, let $\mathbf{x_0}$ denote the initial design used in a two-stage CRM $\mathscr{D}_2$
$\tilde{\mathbf{x}}$ denote a dose sequence generated by this design, that is, $\tilde{x}_i = \mathscr{D}_2(H_i)$
is a predetermined fixed sequence, whereas $\tilde{\mathbf{x}}$
on the toxicity outcomes in the trial. The notion of compatib
can then be expressed as the condition under which $\mathbf{x_0} > \tilde{\mathbf{x}}$
other words, compatibility requires that the initial desig
so to speak, most aggressive) dose escalation plan that is pe
to the trial objective θ
escalation plan when no toxicity is observed; and when there
outcomes, it is undesirable that we should escalate faster t

Another application of Definition 10.1 is to compare two init
used in a two-stage CRM. To illustrate, the dose escalation p
was designed using a two-stage CRM with $N = 33$ patients and $K =$
The study initial design is given by (7.3), which can be equiv
$m_{01} = m_{02} = 3$, $m_{03} = 6$, $m_{04} = 9$, and $m_{05} =$
initial sequence (bottom row) along with three other sequen
the trial. Since larger initial cohort sizes m_{0k}
obvious that the group-of-three design in Table 10.1 is fast
design and the NeuSTART sequence $\mathbf{x_0^{Neu}}$.

Table 10.1 Four examples of initial designs for a trial with $K = 5$ and $N = 33$

Design	m_{01}	m_{02}	m_{03}	m_{04}	m_{05}	$m_{+,4}$
Group-of-three, $\mathbf{x_0^{(3)}}$	3	3	3	3	21	12
Group-of-four, $\mathbf{x_0^{(4)}}$	4	4	4	4	17	16
Increasing cohort size, $\mathbf{x_0^{Inc}}$	2	4	6	8	13	20
NeuSTART, $\mathbf{x_0^{Neu}}$	3	3	6	9	12	21

Furthermore, we can verify that

$$\mathbf{x_0^{Inc}} > \mathbf{x_0^{Neu}}$$

because

$$x_{i,0}^{Inc} \geq x_{i,0}^{Neu} \text{ for all } i$$

with the inequality being strict when $i = 3$,
ing to note that although the group-of-three escalation is f
there exists a sequence (i.e., $\mathbf{x_0^{Inc}}$

of these. In other words, strict ordering does not necessari
quences, and as such, the initial dose sequences constitute

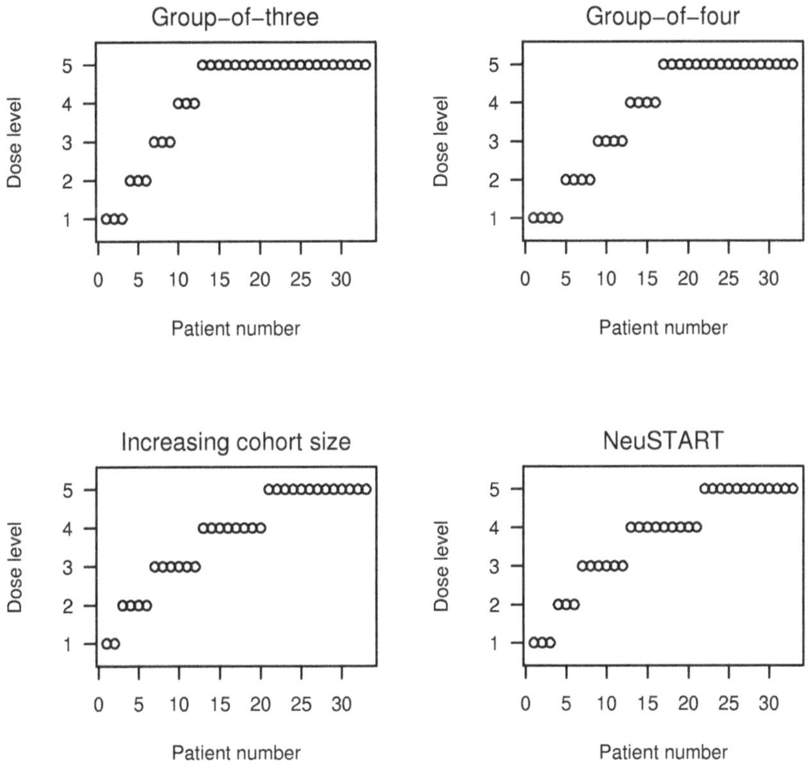

Figure 10.1 Dose level by patient number according to the ini

A simple index to indicate conservatism of an initial sequen
of nontoxic observations required to reach the highest dose, that is,

$$m_{+,K-1} = \sum_{k=1}^{K-1} m_{0k}.$$

A conservative initial design is associated with a large value of $m_{+,K-1}$
let the number of nontoxic observations required by an initi
$j+1$ be denoted as

$$m_{+,j} = \sum_{k=1}^{j} m_{0k}.$$

Definition 10.2 (total ordering). *Let* $\mathbf{x} = \{x_i\}$ *and* $\tilde{\mathbf{x}} = \{\tilde{x}_i\}$
of doses that satisfy (10.1) and (10.2), and $\mathbf{m}$ *and* $\tilde{\mathbf{m}}$
sequence $\mathbf{x}$ *is said to be more aggressive than* $\tilde{\mathbf{x}}$, *denoted by* $\mathbf{x} \succ \tilde{\mathbf{x}}$, *if* $m_{+,J_0} < \tilde{m}$
where $J_0 = \max\{j : m_{+,j} \neq \tilde{m}_{+,j}\}$.

A practical advantage of the relation "$\succ$
can use it to rank dose sequences according to their aggressi
Table 10.1, the sequence $\mathbf{x_0^{Neu}}$ is the least aggressive as it takes $m_{+,4} =$
nontoxic outcomes to escalate to dose level 5, whereas the gr
aggressive with $m_{+,4} = 12$. Overall, we have

$$\mathbf{x_0^{(3)}} \succ \mathbf{x_0^{(4)}} \succ \mathbf{x_0^{Inc}} \succ \mathbf{x_0^{Neu}}.$$

10.3 Building Reference Initial Designs

10.3.1 Coherence-Based Criterion

As stated in Chapter 5, Section 5.2, a one-stage CRM is cohere
CRM is not necessarily so. A logical deduction is that the inc
arises from improper choice of the initial design. Theorem 5
that a cause of incoherence is the use of an incompatible init
incoherence may occur, consider an outcome sequence by a two
$x_{1,0} = \cdots x_{m',0} = d_1$ and $Y_1 = \cdots = Y_{m'-1} = 0$ and $Y_{m'} = 1$ such that $\mathscr{D}_2(H_{m'+1}) \geq$
This sequence is incoherent because an escalation takes place for patient $m' +$
a toxic outcome is observed in patient m'
possible only when m_{01} is set to be sufficiently large, that is, $m_{01} \geq m'$
be avoided by choosing a small m_{01}
design causes incoherence and incompatibility.

To appeal to our intuition, take another instance with a target $\theta = 0$.
quite clear then that an initial design that escalates after
that is, $m_{0k} \equiv 10$, is overconservative. Likewise, an initial sequence with m_{0k}
also appears conservative, though not as much as $m_{0k} \equiv 10$. As we decrease
the initial design seems to be increasingly reflective of the objective $\theta = 0$.
natural question is precisely how small m_{0k}
appropriate. This is the question the coherence (compatibi

Example 10.1.
right panel of the figure shows that the group-of-three initi
Chapter 5, Section 5.2.2 subsequently illustrates that the
is compatible; and so is the group-of-one rule. That is, amon
with constant initial cohort sizes, that is, $m_{0k} = m_0$
$m_{0k} = 1$ or 2. In view of the inclination to be conservative, one may adopt m_{0k}
for $k = 1, \ldots, 4$ in conjunction with this CRM model.

10.3.2 Calibrating Compatible Dose Sequences

As compatibility and the inclination to be conservative are
is reasonable to calibrate the initial design to be the "most
design. There are then two possible approaches to implement
The first approach starts with a compatible initial design, i
dose sequence at each step, and stops iteration once an incom
attained. This calibration approach can be implemented by t

Algorithm 10.1. Base-b compatible benchmark

1. For given θ, K, ν_0, CRM model $F_k(\beta)$, and the prior on β, set $j = \nu_0$ a base (b) initial design $m_{0k} = 0$ for $k < j$, and $= b$ for $k \geq j$.
2. Check the compatibility of the initial design $\{m_{01}, \ldots, m_{0,K-1}, m_{0K}\}^{\dagger}$.
3. If compatibility holds:
   (a) Record this initial design
   (b) Iterate
$$j = \begin{cases} j-1 & \text{if } j > 1 \\ K-1 & \text{if } j = 1 \end{cases}$$
   and then update m_{0j} with $m_{0j} + b$, and repeat Step (2).
4. If compatibility fails to hold in Step 2, stop iteration an initial design in Step 3a.

† The cohort size m_{0K}
is compatible, and can be arbitrarily specified and updated i
as (10.2) is satisfied.

At each iteration, Step 3b moves to test a more conservative i
be proved that once an incompatible sequence is reached in St
dose sequences will be incompatible; see Theorem 10.1 in Sec
last recorded design in Step 3a may be viewed as a conservativ
patible dose sequences, and shall be called a base-b
base-b benchmark satisfies the constraint (10.2).

Example 10.2.
with $b = 1$, we will test dose sequences in the following order:

Iteration	m_{01}	m_{02}	m_{03}	m_{04}	Compatible
1	0	0	1	1	Yes
2	0	1	1	1	Yes
⋮	⋮	⋮	⋮	⋮	⋮
9	2	2	3	3	Yes
10	2	3	3	3	No

We stop at the tenth iteration when we reach an incompatible i

sequence in the ninth iteration, $m_{01} = m_{02} = 2$ and $m_{03} = m_{04} =$
benchmark.

At each iteration, compatibility can be verified using the function cohere
instance, for the initial design at the ninth iteration,

```
> theta <- 0.25
> K <- 5
> p0 <- c(0.05,0.12, 0.25, 0.40, 0.55)
> m0 <- c(2,2,3,3,3) # set m0[5]=3 arbitrarily
> x0 <- rep(1:5, m0)
> foo <- cohere(p0,theta,x0,model="logistic")
> foo$message
[1] "Coherent"
>
```

The base-1 benchmark in Example 10.2 is more conservative th
design obtained in Example 10.1, and may agree more with clin
Ideally, one should seek a compatible initial sequence $\tilde{\mathbf{x}}_0$ such that $\mathbf{x}_0 > \tilde{\mathbf{x}}_0$
compatible $\mathbf{x}_0$. Unfortunately, it is unclear whether such a sequence exist
mainly because of the fact that dose sequences form a partial
displays all possible base-b benchmarks for this CRM model; when $b \geq$
is no compatible initial sequence that starts at $x_{1,0} \leq d_{\nu_0} = d_3$
base-2 design is more conservative than the base-1, but is ne
than the base-8. This confirms the conjecture that the slowes
sequence may not exist due to partial ordering. However, the
most conservative sequence among all, according to the relation $\succ$
dose is a clinical decision, there may be further restrictions on $x_{1,0}$
common to adhere to sequences with $x_{1,0} = d_1$, which implies $m_{01} >$
we will choose between the base-1 and base-2 sequences as the
given CRM model. As a two-stage CRM is motivated by conservat
to choose base-2 benchmark in this example.

Table 10.2 Base-b initial sequences for the CRM model in Example 10.1

b	m_{01}	m_{02}	m_{03}	m_{04}	$m_{+,2}$	$m_{+,3}$	$m_{+,4}$
1	2	2	3	3	4	7	10
2	2	2	4	4	4	8	12
3	0	3	3	3	3	6	9
4	0	4	4	4	4	8	12
5	0	0	5	5	0	5	10
6	0	0	6	6	0	6	12
7	0	0	7	7	0	7	14
8	0	0	8	8	0	8	16

A second calibration approach proceeds in the opposite dire

incompatible initial design, and increase escalation spee
compatible initial design is obtained. An analog of Algorit
implement this calibration approach. However, it can be ver
that this approach will yield the same base-b

10.3.3 Reference Initial Designs for the Logistic Model

This subsection presents some conservative references of i
be used with the logistic model (3.3):

$$F(d_k, \beta) = \frac{\exp\{3 + \exp(\beta) d_k\}}{1 + \exp\{3 + \exp(\beta) d_k\}}$$

in a two-stage Bayesian CRM, where $\beta \sim N(0, 1.34)$
are determined via a specified half-width δ
calibration approach: for a given CRM model, apply Algorith
base-b benchmark for each possible b with $m_{01} >$
conservative sequence according to the total order "$\succ$
did in Section 10.3.2.) The results are shown in Table 10.3 fo
with

- $\theta = 0.10, 0.20, 0.25, 0.33$
- $K = 4, 5, 6, 7$.

The prior MTD ν_0 is set at the median dose level (7.2). The ranges of δ
are chosen in accordance with the optimal δ
$N \geq 30$ and ν_0
conjunction with those tables.

Table 10.3 demonstrates some general trends regarding the e
the reference initial designs. Especially, the initial dos
more aggressive as

- A higher target θ is used, or,
- A larger number K of doses are tested, or,
- The CRM model is specified by a smaller half-width δ.

The first trend is expected; the second trend is also quite cle
comparatively nuanced, but provides an additional conside
in the context of a two-stage CRM. For example, one may be incl
that corresponds to a more conservative reference initial design.

10.4 Practical Issues

10.4.1 Sample Size Constraint

Algorithm 10.1 prescribes a base-b
size m_{0K} at the highest dose. In practice, it may be prudent to set asid
number m^* of observations at the highest level in case of no toxicity, that is,

$$m_{0K} \geq m^*.$$

Table 10.3 Reference initial designs for model (10.4)

θ	K	δ	m_{01}	m_{02}	m_{03}	m_{04}	m_{05}	m_{06}	b
0.10	4	.03	9	9	18	—	—	—	9
		.04	9	18	18	—	—	—	9
	5	.02	5	5	5	10	—	—	5
		.03	5	5	10	10	—	—	5
	6	.02	5	5	5	10	10	—	5
	7	.02	3	3	3	6	6	6	3
0.20	4	.06	4	8	8	—	—	—	4
	5	.05	2	2	4	4	—	—	2
		.06	3	3	6	6	—	—	3
	6	.05	2	2	4	4	4	—	2
	7	.04	2	2	2	4	4	4	2
		.05	1	2	2	2	2	2	1
0.25	4	.06	4	4	8	—	—	—	4
		.07	4	6	6	—	—	—	2
	5	.05	2	2	4	4	—	—	2
		.06	2	2	4	4	—	—	2
		.07	2	2	4	4	—	—	2
	6	.05	2	2	4	4	4	—	2
		.06	2	2	4	4	4	—	2
	7	.05	1	2	2	2	2	2	1
		.07	1	1	2	2	2	2	1
0.33	4	.06	3	3	6	—	—	—	3
		.07	3	3	6	—	—	—	3
		.09	3	3	6	—	—	—	3
	5	.06	1	2	2	2	—	—	1
		.07	1	2	2	2	—	—	1
		.09	1	2	2	2	—	—	1
	6	.06	1	2	2	2	2	—	1
		.07	1	2	2	2	2	—	1
	7	.06	1	1	1	2	2	2	1
		.07	1	1	2	2	2	2	1

The choice of m^* can be partly informed by the target rate θ
set $m^* = 1/\theta$, we will expect to observe one toxic outcome at the highest do
no toxic outcome is observed at the lower doses. To account fo
the actual escalation is slower and more patients will be tre
we may choose $m^* \geq \lambda/\theta$ for some $\lambda >$
choosing λ to be somewhere between 1.5 and 2.0. This recommendation ser
a lower limit of the required sample size; but the final choice
based on the operating characteristics by simulations. At a

is determined by the choice of m^* and the initial design:

$$N \geq m_{+,K-1} + m^*.$$

For example, if we use the reference initial design corresponding to $\theta = 0.25, K$ and $\delta = 0.07$ in Table 10.3, and set $m^* = 1.5/0.25 = 6$, then we may use N
That is, the constraint (10.5) can be used to give a quick asse
the sample size N is adequate; cf,. constraint (7.1) for the one-stage CRM.

In many practical situations, however, the sample size N
and resources available; and a base-b
the available resources can provide. Consider the NeuSTART where $N =$
instance. If we use the reference initial design corresponding to $\theta = 0.10$, K
and $\delta = 0.03$ in Table 10.3, the trial will have enrolled 30 subjects bef
the highest dose and left $m_{05} =$
subjects; the trial may not reach dose level 5 with this initi
observed toxic outcome. Also, this initial design—with $m_{05} =$
(10.2). In other words, a fixed and small sample size N
from being too slow. A pruned base-b benchmark may then be used:

Algorithm 10.2. Pruning a compatible sequence for given N and m^*

1. For given θ, K, ν_0, $F_k(\beta)$, the prior on β
   $\{m_{01}, \ldots, m_{0,K-1}\}$ and set $m_{0K} = N - m_{+,K-1}$.
2. If $m_{0K} < m^*$, set $j = 0$ and proceed with the following steps:
   (a) Iterate
   $$j = \begin{cases} j+1 & \text{if } j < K-1 \\ 1 & \text{if } j = K-1 \end{cases}$$
   and then update m_{0j} with $m_{0j} - 1$ and m_{0K} with $m_{0K} + 1$.
   (b) Stop iteration when $m_{0K} = m^*$.

Example 10.3. Consider $N = 33$ and $m^* =$
initial design $m_{01} = m_{02} = 5$, $m_{03} = m_{04} =$
to Algorithm 10.2 in the following order:

Iteration	m_{01}	m_{02}	m_{03}	m_{04}	m_{05}
0	5	5	10	10	3
1	4	5	10	10	4
2	4	4	10	10	5
⋮	⋮	⋮	⋮	⋮	⋮
8	3	3	8	8	11
9	2	3	8	8	12

The algorithm stops at the ninth iteration, where a compatib
$m_{05} = 12$ is attained.

Step 2a of Algorithm 10.2 reverses Algorithm 10.1 and yields
sequence at each iteration; and hence the pruned sequence wi
than the base-b
pruned sequence will also be compatible.

Pruning may sometimes lead to an unacceptably aggressive in
this be the case, an appropriate approach is to increase the sample size N
are to use (10.5) as a sample size constraint. In practice, it
several pruned compatible designs for a range of Ns and a given m^*
feasible design from a clinical perspective, and confirm the
simulation. In Example 10.3, if $N =$
design with $m_{01} = m_{02} = 5, m_{03} = m_{04} = 10$, and $m_{05} =$
along with the pruned sequence, so that the clinical investi
the tangible trade-off between sample size and conservatism.

10.4.2 Dose Insertion

Dose insertion in phase I trials is a common idea among practi
NeuSTART, for example. The investigators had suggested ins
dose at 7 mg/kg/day between dose tier 3 (6 mg/kg/day) and dose
in case toxicities are observed at dose tier 4. The motivatio
than 6 mg/kg/day (dose tier 3) if a deescalation from dose tie
result. (This suggestion was not implemented in the study fr
perspective; cf., Section 7.2.2.)

From a statistical viewpoint, the cleanest way to handle dos
any post hoc addition. That is, the suggested intermediate d
included for potential testing in the planning stage. In the
7 mg/kg/day is to be used, then one should plan a trial with $K =$
revised dose tiers will then be

Dose tier, k	1	2	3	4	5	6
Lovastatin dose (mg/kg/day)	1	3	6	7	8	10

To reflect that 7 mg/kg/day (dose tier 4) is inserted, we may pr
that skips the dose in a two-stage CRM, that is, the initial cohort size $m_{04} =$
calibration of the initial design can go through the similar
10.1 and 10.2 otherwise.

A simpler alternative is to use the reference initial design
revised dose tiers for NeuSTART, we may use the initial desig
CRM model for $\theta = 0.10, K = 6$, and $\delta = 0.02$ from the table, but set $m_{04} =$
is,

$$m_{01} = m_{02} = m_{03} = 5, m_{04} = 0, m_{05} = 10.$$

Theorem 10.1. *Let* $\mathbf{m_0} = \{m_{0k}\}$ *and* $\tilde{\mathbf{m}}_\mathbf{0} = \{\tilde{m}_{0k}\}$
two sequences $\mathbf{x_0}$ *and* $\tilde{\mathbf{x}}_\mathbf{0}$*, respectively, where* $m_{0k} \leq \tilde{m}_{0k}$
(10.2) are satisfied. If $\tilde{\mathbf{x}}_\mathbf{0}$
$\mathbf{x_0}$ *is also compatible.*

Theorem 10.1 guarantees that (10.6) is a compatible initial
reference design obtained from Table 10.3 is compatible. Fu
impose the sample size constraint $N = 33$ and $m^* =$
be pruned according to Algorithm 10.2:

$$m_{01} = m_{02} = m_{03} = 4, m_{04} = 0, m_{05} = 9, m_{06} = 12.$$

Also as a consequence of Theorem 10.1, the pruned sequence (1

10.5 Case Study: NeuSTART

The NeuSTART adopted a two-stage CRM with $\theta = 0.10, K = 5$, and $\nu_0 =$
empiric (3.2) function

$$F(d_k, \beta) = d_k^{\exp(\beta)}$$

was used to model the toxicity probability at dose level k with labels $d_1 =$
$d_2 = 0.06$, $d_3 = 0.10, d_4 = 0.18$, and $d_5 = 0.30$, where $\beta \sim N(0, 1.34)$
initial dose sequence is given in (7.3). The calibration pro
is described in Section 7.4.2, and is characterized as a tria
the following, we calibrate the initial design by Algorithm
study model (10.8). For comparison purposes, we set $N = 33$ and $m^* =$
in retrospect, we should have set $m^* =$
benchmarks for all possible b so that $m_{01} >$
is most conservative according to the total order $\succ$
with $N = 33$ and $m^* = 12$, we obtain

$$m_{01} = 4, m_{02} = 5, m_{03} = m_{04} = 6, \text{ and } m_{05} = 12.$$

It can be easily verified that the pruned design (10.9) is slow
sequence (7.3).

Table 10.4 Base-b initial sequences for the NeuSTART model

b	m_{01}	m_{02}	m_{03}	m_{04}	$m_{+,2}$	$m_{+,3}$	$m_{+,4}$
1	7	7	8	8	14	22	30
2	6	6	8	8	12	20	28
3	6	6	9	9	12	21	30
4	4	8	8	8	12	20	28
5	5	5	10	10	10	20	30
6	6	6	6	12	12	18	30
7	7	7	7	7	14	21	28

Suppose we opt to adopt the logistic model (10.4) instead of t
NeuSTART. We may choose $\delta = 0$.
suppose we use $\delta = 0$.
as follows:

$$m_{01} = 2, m_{02} = 3, m_{03} = m_{04} = 8, \text{ and } m_{05} = 12.$$

This initial dose sequence is faster than (10.9), but is more original (7.3), that is, (7.3) $\succ$ (10.10) $\succ$ the simulation results of two-stage CRM using these three in along with that of the nonparametric optimal design. The thr comparable accuracy. The logistic model with (10.10) has sl PCS over the five scenarios—0.65 versus 0.63 for the other two varied PCS than the other two. In comparison, the average PCS optimal design is 0.76.

Table 10.5 The distribution of dose selection, average toxi number (OD) of three two-stage CRM for the NeuSTART with $\theta = 0.25, K = 5, \nu_0 = 3$, 33, and $m^* = 12$

Model	$\mathbf{x}_0$	Probability of selecting dose					ATN	OD
Scenario 1:		**.10**	.25	.30	.35	.40		
NP optimal	—	**.92**	.07	.01	.00	.00	—	—
Empiric	(7.3)	**.88**	.11	.01	.00	.00	4.6	7.6
Empiric	(10.9)	**.88**	.11	.01	.00	.00	4.5	7.1
Logistic	(10.10)	**.87**	.12	.01	.00	.00	4.7	8.3
Scenario 2:		.04	**.10**	.25	.30	.35		
NP optimal	—	.23	**.70**	.07	.01	.00	—	—
Empiric	(7.3)	.32	**.53**	.14	.01	.00	3.6	6.8
Empiric	(10.9)	.32	**.52**	.15	.01	.00	3.5	6.2
Logistic	(10.10)	.32	**.54**	.13	.02	.00	3.7	7.2
Scenario 3:		.01	.04	**.10**	.25	.30		
NP optimal	—	.01	.22	**.70**	.07	.01	—	—
Empiric	(7.3)	.02	.27	**.56**	.14	.01	3.1	5.4
Empiric	(10.9)	.02	.26	**.56**	.13	.02	3.0	5.1
Logistic	(10.10)	.02	.26	**.55**	.14	.03	3.3	5.8
Scenario 4:		.01	.01	.04	**.10**	.25		
NP optimal	—	.00	.01	.22	**.70**	.08	—	—
Empiric	(7.3)	.00	.03	.25	**.56**	.16	2.4	2.9
Empiric	(10.9)	.00	.03	.26	**.51**	.19	2.3	3.3
Logistic	(10.10)	.00	.03	.25	**.49**	.23	2.6	4.3
Scenario 5:		.01	.01	.01	.04	**.10**		
NP optimal	—	.00	.00	.01	.22	**.77**	—	—
Empiric	(7.3)	.00	.01	.05	.28	**.66**	1.4	0.0
Empiric	(10.9)	.00	.01	.06	.26	**.67**	1.3	0.0
Logistic	(10.10)	.00	.01	.03	.18	**.78**	1.5	0.0

Note: The results for the nonparametric optimal design (NP o references. Numbers associated with the MTD are in bold.

The logistic model causes more toxicity and overdose than th difference is small. All three CRM designs lead to much reduc

compared to balanced randomization. This illustrates the c
two-stage CRM design.

As we have emphasized in the previous chapters, the CRM desig
the automated calibration approach give comparable operat
original labor-intensive NeuSTART design. In particular,
start to calibrate and prune a reasonable initial design for a two-stage CRM.

10.6 Exercises and Further Results

Exercise 10.1. Verify that $\mathbf{x} > \tilde{\mathbf{x}}$ implies $\mathbf{x} \succ \tilde{\mathbf{x}}$.

Exercise 10.2.
model (10.4) with $\delta = 0.02$, give an initial sequence under the constraint N
and $m^* = 12$. Use simulation to compare this two-stage CRM with the des
Table 10.5.

Exercise 10.3. Using Theorem 10.1, verify that pruning using Algorithm 10.
result in a compatible sequence.

Part III

CRM and Beyond

Chapter 11

The Time-to-Event CRM

11.1 Introduction

In the bortezomib trial, each patient would receive up to six chemotherapy, and would thus be at risk of toxicity througho period. If each patient was to be followed for 126 days before made for the next patient, the trial duration would be imprac hand, the common phase I practice would usually count toxici the first cycle, and would underestimate the incidence of adv might occur at a later cycle. To deal with the prospect of *late toxicity* trial adopted a variant of the CRM, called the time-to-event would allow continual accrual throughout the trial while us endpoint as basis of dose escalation.

The basic idea of the TITE-CRM (read "tight C-R-M") is to esti toxicity curve based on the current toxicity status of all pa have not received a full six-cycle treatment. Section 11.2 m and presents the mathematical framework of the method. Sect numerical example to illustrate the implementation of the m discusses the practical issue of enrollment scheduling. Th method are outlined in Section 11.5. Two-stage design with t described in Section 11.6, and applied to the Poly E trial int The chapter is concluded in Section 11.7 with a brief review o TITE-CRM in the literature, and alternatives to deal with late toxicities.

11.2 The Basic Approach

11.2.1 A Weighted Likelihood

A one-stage Bayesian TITE-CRM adopts the same dose assignme regular CRM, and treats a newly enrolled patient at the model The main difference is that the TITE-CRM uses information fr patients in addition to completely followed patients. For e patient has been on treatment in the bortezomib trial for five of toxicity. Although we have not observed the patient's out a five-cycle toxicity-free follow-up speaks for the safety o receives, and in some sense may be counted as five-sixths of a n

This idea is made concrete through the use of a weighted likel
parameter, defined as

$$L_n(\beta;\mathbf{w}) = \prod_{i=1}^{n} \{w_{i,n+1}F(x_i,\beta)\}^{Y_{i,n+1}} \{1 - w_{i,n+1}F(x_i,\beta)\}^{1-Y_{i,n+1}}$$

where $Y_{i,n+1}$ and $w_{i,n+1}$ are, respectively, the toxicity indicator for patient i
weight assigned to this observation just prior to the entry of patient $n+$
$Y_{i,n+1}$ is an increasing process in n and $Y_{i,n} = 1$ implies $Y_{i,n'} = 1$ for every n'
Generally, the eventual toxicity status $Y_{i,N+1}$ is the same as Y_i
(Recall N denotes the sample size of the trial.) As patient $n+$
for the trial, the parameter β is estimated by

$$\hat{\beta}_n^w = \frac{\int_{-\infty}^{\infty} \beta L_n(\beta;\mathbf{w})\, dG(\beta)}{\int_{-\infty}^{\infty} L_n(\beta;\mathbf{w})\, dG(\beta)}$$

so that the patient will receive the model-based MTD estimate

$$x_{n+1} = \arg\min_{d_k} |F(d_k, \hat{\beta}_n^w) - \theta|.$$

Alternatively, one may estimate β by maximizing the weighted likelihood. Let

$$\tilde{\beta}_n^w = \arg\max_{\beta} L_n(\beta;\mathbf{w})$$

denote the maximum weighted likelihood estimate, which exists if $Y_{j,n+1} =$
$Y_{j',n+1} = 1$ for some $j, j' \leq n$
likelihood TITE-CRM is used, as is the case for the regular CR
erogeneity in the toxicity outcomes among patients, the dos
$n+1$ will be:

$$x_{n+1} = \arg\min_{d_k} |F(d_k, \tilde{\beta}_n^w) - \theta|.$$

Before there is any toxic outcome in the trial, one will eithe
according to a predetermined dose sequence $\{x_{i,0}\}$
tion 11.6 for further discussion on the specification of two-stage TITE-CRM.

11.2.2 Weight Functions

The TITE-CRM incorporates the times-to-event of the patien
in the weighted likelihood. Intuitively, the weight $w_{i,n+1}$
information contributed to the likelihood by patient i
increases with the patient's length of follow-up and that $w_{i,n+1} =$
a complete follow-up.

In the case where a new patient is enrolled only when all curre
been completely followed, the weighted likelihood reduces
likelihood, and the TITE-CRM to the regular CRM. More genera
$w_{i,n+1}$s, are determined in a manner that is independent of β

with the toxic outcomes can be factored out as a proportional
likelihood and do not affect the estimation of β
next patient. Thus, we may set $w_{i,n+1} = 1$ if patient i
It is indeed intuitive to regard an observed toxic outcome as
irrespective of the length of follow-up.

To study the weighting scheme more rigorously, let T_i
of patient i, and T
observation window. The observation window should be presp
protocol, so that a patient is said to experience a toxicity (i.e., $Y_i =$
if $T_i \leq T$. For example, the bortezomib trial has an observation window of $T =$
days. Then by the definition of conditional probability, for $t \leq T$, we can write

$$\Pr(T_i \leq t \mid x_i) = \Pr(T_i \leq t \mid T_i \leq T, x_i)\Pr(T_i \leq T \mid x_i).$$

In the regular CRM, the marginal distribution $\Pr(T_i \leq T \mid x_i)$
the dose–toxicity function $F(x_i, \beta)$
$\Pr(T_i \leq t \mid T_i \leq T, x_i)$ is approximated by weight function $w(t; T, x_i)$
likelihood function based on current toxicity status can be written as

$$\prod_{i=1}^{n} \{\Pr(T_i \leq C_{i,n+1} \mid x_i)\}^{Y_{i,n+1}} \{1 - \Pr(T_i \leq C_{i,n+1} \mid x_i)\}^{1-Y_{i,n+1}}$$
$$= \prod_{i=1}^{n} \{w(C_{i,n+1}; T, x_i)F(x_i, \beta)\}^{Y_{i,n+1}} \{1 - w(C_{i,n+1}; T, x_i)F(x_i, \beta)\}^{1-Y_{i,n+}}$$

where $C_{i,n+1}$ is the length of follow-up of patient i
patient $n+1$. Technically, this likelihood is valid if the length of follow-up $C_{i,n}$
independent of the time-to-toxicity T_i for all i
trials, where the censoring times are essentially administ
the arrival times of new patients. Now, by comparing the weig
and the current status likelihood (11.5), we see that the weight $w_{i,n+1}$
may be viewed as an estimate of the weight function $w(C_{i,n+1}; T, x_i)$
weighted likelihood may thus be viewed as an approximate lik
current status data.

A practical implication of the conditioning argument (11.4
modeling and time-to-toxicity modeling can be done separat
approach for $F_k(\beta)$
previous chapters. To model the conditional time-to-toxic
and Chappell [20] consider a simple linear weighting scheme
proportional to the length of follow-up. With respect to the
probabilistic interpretation of $w(t; T, x_i)$, that is,

$$w(0; T, x_i) = 0 \text{ and } w(t; T, x_i) = 1 \text{ for } t \geq T,$$

the linear weight function is defined as

$$w(t; T, x_i) = \min\left(\frac{t}{T}, 1\right).$$

The linear weight (11.6) appears to be an oversimplistic est
distribution $\Pr(T_i \leq t \mid T_i \leq T, x_i)$
of the distribution and assumes identical shape for all dose
may use an adaptive weighting scheme in which the shape of the
is altered based on the accrued observations. In particular, with n
Cheung and Chappell [20] consider

$$w_n(t;T,x_i) = \frac{\kappa(t)}{z+1} + \frac{1}{z+1}\left\{\frac{t - t_{(\kappa(t))}}{t_{(\kappa(t)+1)} - t_{(\kappa(t))}}\right\} \text{ for } t \leq T,$$

where z is the total number of toxic outcomes, the times-to-toxicity $t_{(j)}$
such that $0 \equiv t_{(0)} < t_{(1)} \leq \cdots \leq t_{(z)} < t_{(z+1)} \equiv T$, and $\kappa(t) = \max\{j \in [0,z] : t \geq t_{(}$
When there is no toxic outcome, that is, $z =$
identical to the linear weight function.

The functional form of w_n
the same weight for patients at different doses as long as the
of follow-up. Cheung [17] studies dose-adjusted weight tha
distribution of T_i given x_i
higher doses, that is, stochastic ordering of the time-to-t

$$\Pr(T_i \leq t \mid x_i) \leq \Pr(T_{i'} \leq t \mid x_{i'}) \text{ for } x_i < x_{i'}.$$

This regression model is fairly general. On the other hand, t
dose-adjusted weights is complicated and requires the iter
algorithm [3].In addition, a moderate number of observatio
to precisely estimate the dose-adjusted conditional distribution $\Pr(T_i \leq t \mid T_i \leq T,$
To be more parsimonious, one may estimate the conditional distribution of T_i
given dose by parametric regression such as the accelerated
a Weibull distribution [20] and the cure model [12].

To summarize, the linear weight (11.6) is found to be adequat
Using an adaptive weighting scheme generally does not impro
the TITE-CRM, but rather affects the in-trial allocation in
When the late-onset tendency of toxicity is prominent (e.g.
adaptive weights such as (11.7) will reduce the exposure of p
doses. Additionally, an adaptive weight scheme should be us
we expect nonnegligible amount of patient dropout; see Theo
Section 11.5.

11.2.3 Individual Toxicity Risks

A derived concept from the time-to-toxicity distribution i
Precisely, define the toxicity risk for patient i at time t, for $t < T$, as

$$r_i(t) = \Pr(T_i \leq T \mid T_i > t, x_i).$$

The toxicity risk (11.8) measures the likelihood of toxicity for patient i
maining treatment period, provided that the patient is toxicity-free up to time t

algebraic manipulation will further give

$$r_i(t) = \frac{\{1-\Pr(T_i \le t \mid T_i \le T, x_i)\}\Pr(T_i \le T \mid x_i)}{1-\Pr(T_i \le t \mid x_i)},$$

from which it is clear that $r_i(0) = \Pr(T_i \le T \mid x_i)$, and $r_i(t) \to 0$ as $t \to T$
it can be shown that $r_i(t)$ is a decreasing function of t
has been followed without observing toxicity, the lower ris
has.

As in (11.4), we can model $r_i(t)$ through the marginal model $F(x_i, \beta)$
weight function $w(t; T, x_i)$. In particular, just prior to the entry of patient $n+$
risk of patient i, $r_i(C_{i,n+1})$, can be approximated by

$$r_{i,n+1} = \frac{(1-w_{i,n+1})F(x_i, \hat{\beta}_{comp})}{1-w_{i,n+1}F(x_i, \hat{\beta}_{comp})}$$

where $\hat{\beta}_{comp}$ is the posterior mean of β
by $H_{comp} = \{(x_j, Y_{j,n+1}) : w_{j,n+1} = 1\}$. In words, the complete data history
contains the observations from patients either with comple
toxic outcome. The risk estimate (11.9) is applied to patien
the interim. If a toxic outcome has already been observed in patient i
a certainty, and we will set $r_{i,n+1} = 1$.

11.3 Numerical Illustration

11.3.1 The Bortezomib Trial

To recall, the bortezomib trial used the TITE-CRM for dose es
toxicity relationship was modeled via the empiric function
$d_1 = 0.05, d_2 = 0.12, d_3 = 0.25, d_4 = 0.40$, and $d_5 = 0.55$ and $\beta \sim N(0, 1.34)$
calibration details can be found in Chapter 7, Section 7.4.1.

To incorporate the observed times-to-toxicity into the estimation of β
weight function (11.6) was used to compute the weighted like
Table 11.1 displays the current status of the first 4 patients
date of patient 5. Since the observation window in the trial i
73-day follow-up has a weight of $73/126 = 0.$
can be computed in the same manner. Since there has been no tox
patients receive dose level 3, the weighted likelihood is

$$\prod_{i=1}^{4}\{1 - w_{i,5} \times 0.25^{\exp(\beta)}\}$$

where $w_{i,5}$s are given in Table 11.1. As a result, the posterior mean $\hat{\beta}_5^w = 0.$
the recommended dose by the TITE-CRM at this time is dose leve
subsection for the R code that produces these numerical results.

The toxicity risks for the patients are estimated using (11.

who receives dose 3 and has an assigned weight 0.58. His risk f
remaining treatment period is approximated by

$$r_{1,5} = \frac{(1 - w_{1,5})\, F(d_3, \hat{\beta}_{comp})}{1 - w_{1,5} F(d_3, \hat{\beta}_{comp})} = \frac{(1-0.58) \times 0.25}{1 - 0.58 \times 0.25} = 0.123,$$

where $\hat{\beta}_{comp} = 0$; the prior mean of β
followed at this time and H_{comp} is an empty set.

Table 11.1 The current status of the first 4 patients in the bor
patient 5 (April 16, 2004), and the estimates of their individual risks of toxicity

i	x_i	Entry date	$Y_{i,5}$	$C_{i,5}$	$w_{i,5}$	$r_{i,5}$
1	3	February 3, 2004	0	73	0.58	0.123
2	3	February 10, 2004	0	66	0.52	0.137
3	3	March 12, 2004	0	35	0.28	0.194
4	3	March 19, 2004	0	28	0.22	0.206

11.3.2 Implementation in R

The TITE-CRM is implemented by the R function `titecrm`

```
> ### Determine dose for patient 5 in the bortezomib trial
> p0 <- c(0.05,0.12,0.25,0.40,0.55) # initial guesses
> theta <- 0.25
> y <- c(0,0,0,0)              # current toxicity status
> x <- c(3,3,3,3)              # dose level
> cc <- c(73,66,35,28)         # lengths of follow-up (in days)
> wt <- cc / 126               # linear weights
> foo <- titecrm(p0,theta,y,x,weights=wt)
> foo$estimate                 # posterior mean of beta
[1] 0.4907791
> round(foo$ptox,digits=2) # posterior toxicity probability
[1] 0.01 0.05 0.14 0.27 0.42
> foo$mtd                      # model-based MTD
[1] 4
>
```

There are various ways to enter the weight information to the function `titecrm`
most straightforward approach is to use the argument `weight`
Alternatively, one could enter the patients' lengths of fol
`followup`, along with the observation window via the argument `obswin`
weights of the patients will be computed according to the lin
adaptive weight (11.7).

```
> ### The following gives an equivalent object to 'foo'
```

```
> foo2 <- titecrm(p0,theta,y,x,followup=cc,obswin=126)
> foo2$weights
[1] 0.5793651 0.5238095 0.2777778 0.2222222
> foo2$estimate          # posterior mean of beta
[1] 0.4907791
>
```

The default weighting scheme used by titecrm
the argument scheme
will be used. In this example, since there is no toxic outcome
will be identical to the linear weights.

In addition, the R function titesim
under a given dose–toxicity curve. Exercise 11.1 demonstra
function as an example.

11.4 Enrollment Scheduling

11.4.1 Patient Accrual

The data in Table 11.1 led to the first escalation in the bortez
first 4 patients had been followed for a total of 202 patient-d
toxicity; cf., Figure 1.1. To compare, a CRM with the same dos
prior would escalate dose after a complete nontoxic outcome
This illustrates that the TITE-CRM is a reasonably cautious
and can be applied with a fully sequential enrollment plan wi
enrollment.

On the other hand, one will need to exercise caution if the pat
rapid. Figure 11.1 shows the operating characteristics of t
the bortezomib trial under the five dose–toxicity scenarios
patient accrual rate ξ. In each simulated trial, the interpatient arrival times we
so that ξ patients would be enrolled per observation window; for example, for ξ
with an observation window of 126 days, patient 1 would be enr
patient 2 day 84, patient 3 day 126, and so on. The regular CRM w
corresponding to $\xi =$
patients were completely followed. The conditional unifor
erate the times-to-toxicity in the simulations [20]: we firs
a toxic response; if so, we would generate a number on the inte
as the time-to-toxicity.

The figure shows general trends that the PCS declines as ξ
toxicity and overdose numbers increase. These trends indic
of erroneous escalations due to inadequate follow-up of ind
patients are toxicity-free after very limited follow-up, t
weighted likelihood may occasion an escalation. Thus, the T
an increased number of patients at toxic doses before any toxicity is observed.

However, the impacts on the operating characteristics are g
the reduction in trial duration shown in Figure 11.1d. The bo

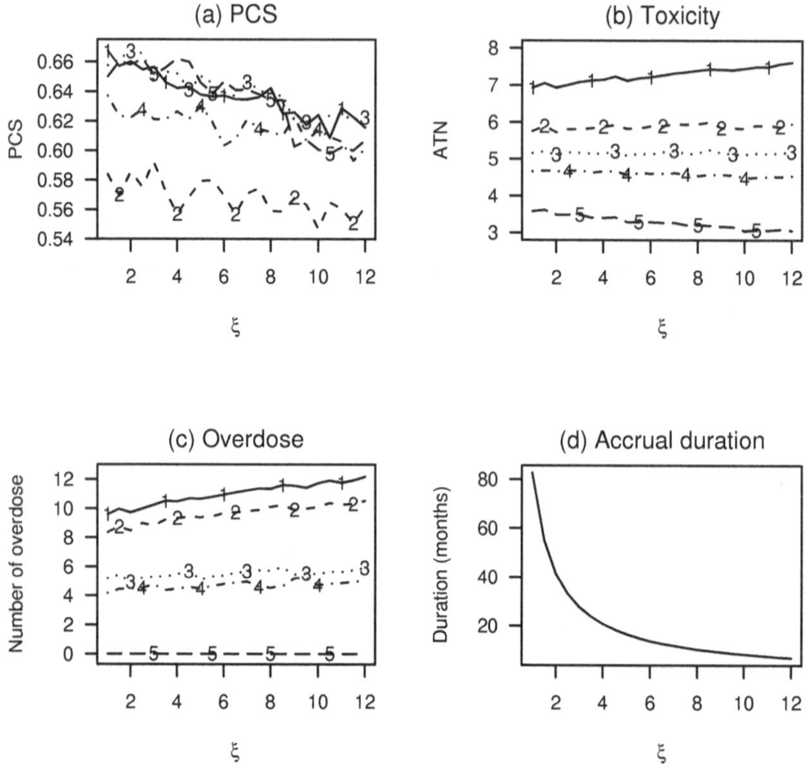

Figure 11.1 Operating characteristics of the TITE-CRM against patient accrual rate ξ
the dose–toxicity scenarios in Table 7.1. The numbers in figu
simulation scenarios.

days to enroll 20 patients (i.e., roughly 6.2 patients per 12
retrospectively compare the TITE-CRM with $\xi = 6.5$ and the CRM ($\xi =$
drops by 1–3 percentage points, the increase in the average n
greater than 0.4 per trial, and the number of overdose under s
0.7–1.5 per trial. In comparison, there is a sixfold differe
the TITE-CRM reduces the accrual period from 7 years to 14 mon
Figure 11.1d. Also, if a group accrual CRM with size 3 is used,
for $N = 20$ will be roughly 3 years if we assume patients are immediate
While this reduction is substantial, a 3-year long phase I tr
size may prove infeasible in many clinical settings where th
and practical considerations.

11.4.2 Interim Suspensions

When patient availability is quick relative to the observat
the study temporarily as a safety measure, so that dose decis
can be based on adequate follow-up of the current patients. I
sion decisions may be made through the mechanism with a data s
ing committee (DSMC). However, it will be reassuring if the i
the monitoring rules in the study protocol. This section pre
ing approach used in conjunction with the TITE-CRM. In brief
invoke accrual suspension if there is a nonnegligible likel
will treat the next patient at an overdose as a result of not ha
data. This process entails the enumeration of all possible o
all the patients be completely followed, and the quantificat
each sequence. The concrete steps are carried out as follows:

Algorithm 11.1. Interim risk monitoring in the TITE-CRM

At an interim with n enrolled patients, let x_{n+1}
the TITE-CRM based on the observed data. Then the monitoring
assess the risk of x_{n+1} using the following steps:

1. Enumerate all possible eventual outcome configurations g
   Each configuration, denoted by $\mathbf{y^f_{n+1}} = (y^f_{1,n+1}, \ldots y^f_{n,n+1})$
   set.
2. For each full outcome set, evaluate the recommended dose, denoted as x^f_{n+1}
3. Approximate the *risk likelihood*
   the complete data H_{comp}.
4. Accumulate the risk likelihoods of all full outcome sets with $x^f_{n+1} < x_{n+1}$
   accumulation measures the likelihood of selecting a lower dose than x_{n+1}
   all patients are completely followed. Thus, suspend the tri
   "large" such as when ≥ 0.50.

To illustrate these steps, consider the bortezomib data in T
were partially followed with no toxic outcome on April 16, 2004, there were 2^4
possible full outcome sets. Table 11.2 lists each of the 16 ou
their recommended dose x^f_5
and 2 of Algorithm 11.1.

To calculate the likelihood of each full outcome (Step 3), we
individual risks of toxicity as shown in Table 11.1. With the
likelihood of each full outcome is computed as

$$\prod_{i=1}^{n} (r_{i,n+1})^{y^f_{i,n+1}} (1 - r_{i,n+1})^{1-y^f_{i,n+1}}.$$

For example, for the outcome $y^f_{1,5} = y^f_{2,5} = y^f_{3,5} = y^f_{4,5} =$

Table 11.2 All possible full outcome sets in the first 4 patien
April 16, 2004 (see Table 11.1), and their corresponding rec
of occurrence

$y^f_{1,5}$	$y^f_{2,5}$	$y^f_{3,5}$	$y^f_{4,5}$	x^f_5	Likelihood
0	0	0	0	5	0.484
1	0	0	0	3	0.068
0	1	0	0	3	0.077
1	1	0	0	1	0.011
0	0	1	0	3	0.117
1	0	1	0	1	0.016
0	1	1	0	1	0.019
1	1	1	0	1	0.003
0	0	0	1	3	0.126
1	0	0	1	1	0.018
0	1	0	1	1	0.020
1	1	0	1	1	0.003
0	0	1	1	1	0.030
1	0	1	1	1	0.004
0	1	1	1	1	0.005
1	1	1	1	1	0.001

likelihood is

$$(1-0.123)\times(1-0.137)\times(1-0.194)\times(1-0.206)=0.484.$$

The recommended x^f_5
for this particular outcome, the recommended doses based on
lower than what the TITE-CRM would choose (i.e., dose level 4
thus seems quite likely that the TITE-CRM is recommending a d
MTD estimate should we wait long enough for all patients to be
The likelihood of this event is $0.516 = 1 - 0.$
temporary accrual suspension.

While we did not implement Algorithm 11.1 or other monitorin
bortezomib trial, this type of risk calculation could have b
the protocol. Generally, the calculations can be carried ou
when the TITE-CRM recommends an escalation from the current
DSMC review purposes, it will be useful to submit the risk lik
safety information, along with the method calculation. Fur
suspension threshold for the risk likelihood in Step 4 of the
decisions during the trial.

Bekele et al. [8] propose a trial monitoring approach that us
risks of toxicity at prospective doses for future patients.
Algorithm 11.1: accrual will be suspended if the predicted r
Obtaining Bayesian predicted risks, however, requires spe

Markov Chain Monte Carlo. Algorithm 11.1 is comparatively l
and can be easily programmed using the functions in the 'dfcrm' package.

11.5 Theoretical Properties

11.5.1 Real-Time Formulation

To study the theoretical properties of the TITE-CRM, it is us
to reconcile calendar time, t, and an individual patient's time, s, where $s =$
the time the patient enters the trial. Let e_i
enters the trial. For any calendar time t, the patient's length of follow-up is $C_i($
$\min\{\max(t - e_i, 0), T\}$. The weighted likelihood (11.1) can then be rewritten as

$$L(\beta, t) = \prod_{i=1}^{N(t)} \{w_i(t)F(x_i,\beta)\}^{Y_i(t)} \{1 - w_i(t)F(x_i,\beta)\}^{1-Y_i(t)}$$

where $Y_i(t) = I\{T_i \leq C_i(t)\}$ and $w_i(t) = w\{C_i(t); T, x_i\}$
toxicity status and the weight of patient i, and $N(t)$
prior to calendar time t.

11.5.2 Real-Time Coherence

In summary, a coherent design does not allow escalation righ
and deescalation after a nontoxic outcome. Real-time versi
for the situations where toxicity may occur randomly over an
window while patients are enrolled continually in a stagger
in the bortezomib trial. Naturally, a real-time version of c
does not allow deescalation from a time point t to another time point $t + h$
toxicity is reported during the time period $[t, t + h)$
design $\mathscr{D}$ is coherent in deescalation if, with probability one,

$$P_{\mathscr{D}}\{U(t,h) < 0 \mid Y_i(t+h) - Y_i(t) = 0 \text{ for all } i\} =$$

where $U(t,h) = \mathscr{D}(H_{t+h}) - \mathscr{D}(H_t)$, and $H_t \equiv \{(x_i, Y_i(t), T_i \wedge C_i(t)) : i \leq N(t)\}$
entire history of observations just prior to time t.

A design is called real-time coherent in escalation if it doe
instant immediately after a toxic observation: with probability one,

$$\lim_{h \to 0} P_{\mathscr{D}}\{U(t,h) > 0 \mid Y_i(t+h) - Y_i(t) = 1 \text{ for some } i\} = 0.$$

The conditions given by (11.11) and (11.12) imply (5.2) and (
the toxicity outcomes can be evaluated instantaneously. Al
to interpret (11.12), the notion of real-time coherence pro
to evaluate a real-time dose finding method.

Theorem 11.1. *If the weight function is defined such that $w_i(t)$*
nondecreasing in t for all i, then the one-stage Bayesian TIT
coherent.

It is easy to verify that the linear weight (11.6) satisfies th
the theorem. For other adaptive weight functions such as (11
a restriction that each patient's weight is not to decrease a
proof of Theorem 11.1 is given in Cheung [16].

11.5.3 Consistency

Cheung and Chappell [20] and Cheung [17] study the asymptoti
the maximum likelihood TITE-CRM under the regularity condi
toxicity model stated in Chapter 4, Section 4.3, with additi
the weight function and the patient accrual process. These a
modified to suit the Bayesian TITE-CRM as in Chapter 5, Sectio
in the following:

Theorem 11.2. *Suppose that the conditions in Theorem 5.3 hold. If*

$$\sum_{i=1}^{N(t)} I\{C_i(t) < T\}/N(t) \rightarrow$$

as $t \rightarrow \infty$ and the weight function does not depend on β
select dose d_l eventually with probability one, and $l = \nu$ is the true MTD.

Theorem 11.2 requires the same set of assumptions in regard t
model $F_k(\beta)$ as in the CRM. A practical implication is that calibration of
toxicity model for the TITE-CRM may be done using the same tec
CRM. For time-to-event modeling, Theorem 11.2 allows a broa
functions, provided that the number of incomplete observat
pared to the number of enrolled patients. The assumption (11
situations where we expect a certain level of patient dropou
mate concern for trials with an extended follow-up period.

Theorem 11.3. *Suppose that the conditions in Theorem 5.3 hold. If the proce*
of patient i is conditionally independent of the eventual outcome Y_i
be observed) given $Y_i(t)$, and that

$$\sup_{s\in[0,T]} \left| w_{N(t)}(s;T,x_i) - Pr(T_i \leq s \mid T_i \leq T, x_i) \right| \rightarrow$$

as $t \rightarrow \infty$, the TITE-CRM will select dose d_l
$l = \nu$ is the true MTD.

The assumption in Theorem 11.3 amounts to a missing-at-rand
dropout pattern [91]. While MAR is an untestable assumption
itly used in dose finding trials. The assumption is reasonabl
one condition: if dose withdrawal is due to adverse events ca
event should be defined as dose limiting even though the adver

have reached the severity level of a dose-limiting toxicity
gible amount of patient dropout, the consistency requireme
in Theorem 11.3 becomes stricter than that in Theorem 11.2. I
ear weight (11.6) does not generally meet the condition in Th
the adaptive weight function (11.7) when the conditional distribution of T_i
on dose x_i. However, with the typical small-to-moderate sample size i
studies, the data may prove to be too sparse for any dose-adju
Extensive simulation shows that the use of adaptive functio
to a dose-adjusted scheme when there is a moderate level (30%
though the assumption (11.14) is not met by the function (11.7) generally [17].

11.6 Two-Stage Design

11.6.1 Waiting Window

Like a two-stage CRM, a two-stage TITE-CRM is implemented in
prespecified dose sequence $x_{1,0} \leq \cdots \leq x_{N,0}$
the initial design includes the minimum length of waiting ti
window τ, between cohorts before an escalation may take place in the i
Thus, a two-stage TITE-CRM avoids aggressive dose assignme
it starts the trial at a low dose. Second, the waiting window a
escalation.

Figure 11.2 shows a simulated trial of size 30 using a two-sta
a group-of-three initial escalation rule starting at the lowest dose. There are K
dose levels, and the trial objective is to identify a dose with a $\theta = 0.$
probability during a 6-month period. In the simulation, the
fixed at a rate of one patient per month (i.e., $\xi =$
during stage 1 where a 4-month waiting window is imposed betw
escalation is to take place (i.e., $\tau =$
at 4 months after patient 3. There is another 4-month delay af
patient 6. This delay allows the observation of the toxic out
enrolling patient 7; this toxic observation then triggers t
based TITE-CRM, and patients are enrolled at the rate of 1 per
delay. As a result of the waiting window in the initial stage,
recruit 30 subjects. In comparison, an uninterrupted accru
months to enroll.

This example illustrates the trade-off between timeliness
through accrual suspensions. As the risk calculations in Al
prior information before any toxicity is seen, the waiting w
in a two-stage design. However, the window τ
input parameter that represents a duration in which toxicit
probable.

Theorem 11.4. *Suppose the initial design* $x_{1,0} \leq \cdots \leq x_{N,0}$
two-stage CRM is coherent per Theorem 5.1. A two-stage TITE-

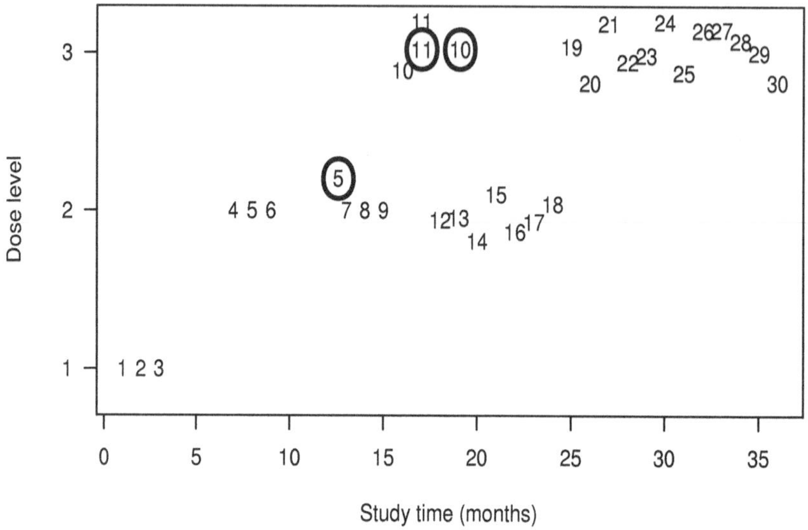

Figure 11.2 A simulation of a two-stage TITE-CRM trial using
Section 11.6.2. Each number indicates a patient: A number ma
circled number indicates the time when a toxicity occurs. Ve
are jittered for visual clarity.

dose–toxicity model and initial design sequence is real-ti
the waiting window τ.

Theorem 11.4 implies that we may specify the initial dose sequence $\{x_{i,0}\}$
guidelines in Chapter 10. The sketch of the proof of Theorem 1
Exercise 11.3.

11.6.2 Case Study: The Poly E Trial

Chemoprevention involves the use of specific agents to rever
carcinogenic progression to invasive cancer [65]. Chemopr
many diverse populations: healthy individuals, individua
cancer survivors who are at risk for recurrence or new primar
populations are generally healthier than cancer patients.
a long process, chemoprevention will be given over an extend
toxicities may be become intolerable, and long-term effect
it is important to allow an adequate follow-up when evaluati
agent. These concerns are exemplified in the Poly E trial intr

in Chapter 2. This dose-escalation study tested $K =$
600, or 800 mg taken orally per day over a 6-month period, in $N =$
history of breast cancer. The lowest dose level 400 mg was cho
preclinical and clinical safety data, and would be the start
$x_1 = d_1$. The objective was to identify a dose causing $\theta = 0.$
that was to occur during the 6-month treatment period. In thi
higher grade toxicity would be counted as dose limiting; thi
commonly used grade 3 dose-limiting threshold in chemother
lower dose-limiting threshold was used for feasibility rea
as headache over an extended period might prove intolerable
at risk but lived without cancer, and thus prevent the subjec
compliance with the treatment.

The TITE-CRM was used to determine dose escalation. The line
was used so that the weight of an observation was proportiona
follow-up. For dose–toxicity modeling, the empiric function

$$F(d_k, \beta) = d_k^{\exp(\beta)}$$

was used with $\beta \sim N(0, 1.34)$ and dose labels $d_1 = 0.15$, $d_2 = 0.25$, and $d_3 =$
The choice of this dose–toxicity model was made by trial-and
manner as in the bortezomib trial and the NeuSTART; see Secti
this model, the prior MTD $\nu_0 = 2$ and $x_1 < d_{\nu_0}$
with a group-of-five initial design, that is,

$$x_{1,0} = \cdots = x_{5,0} = d_1,\ x_{6,0} = \cdots = x_{10,0} = d_2,\ x_{11,0} = \cdots x_{30,0} = d_3.$$

This initial design can be shown to be incompatible with mode
cause incoherent escalation: The concept of coherence had n
at the time the Poly E trial was planned; rather, it was decide
committee. However, there was a restriction that no escalat
the previous patient had a toxic outcome. In other words, no c
allowed.

In this section, we will focus on the use of model (11.15) in th
E trial, and examine the effects of the initial design—the dose sequence $\{x_{i,0}\}$
the waiting window τ
Now, applying Algorithm 10.1, we obtain the group-of-three
a conservative base-b benchmark, that is,

$$x_{1,0} = x_{2,0} = x_{3,0} = d_1,\ x_{4,0} = x_{5,0} = x_{6,0} = d_2,\ x_{7,0} = \cdots = x_{30,0} = d_3.$$

Therefore, for the initial design, we will consider

- Initial dose sequence $x_{i,0}$s: group-of-three versus group-of-five
- Waiting window τ : $1, 2, \ldots, 6$ months.

For the two-stage design with group-of-five initial sequenc
by restriction.

Figure 11.3 shows the PCS and the safety indices of the TITE-C

scenario with $p_1 = 0.10, p_2 = 0.25$, and $p_3 = 0.$
and level 3 an overdose. The results are based on a fixed enroll
$\xi = 6$ patients per observation window (6 months). Overall, the i
dose sequence and the waiting window τ
be noted. First, the group-of-three initial design induces
and overdose than the group-of-five design; this is expected
calates faster than the latter. Second, the toxicity indices decline as τ
is also expected because erroneous escalations may be avoid
sions based on longer follow-up data. However, all design sp
represent improved safety upon balanced randomization, wh
patients at an overdose under this dose–toxicity scenario r
7.5 toxic outcomes. This being said, if the group-of-three s
be prudent to be cautious by setting $\tau =$
will add about 3 months to the trial duration.

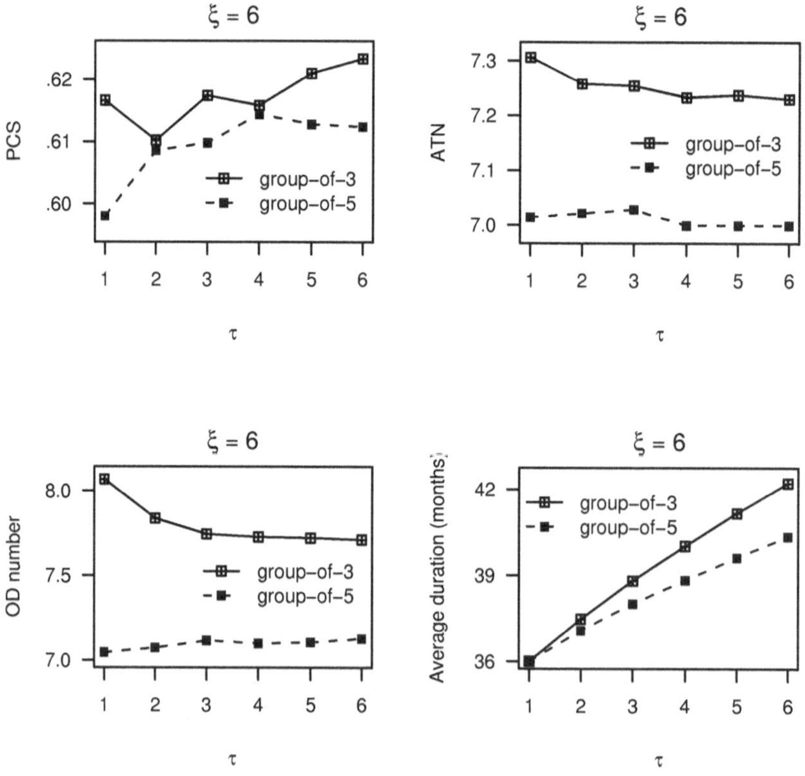

Figure 11.3 Operating characteristics of the two-stage TIT
various initial design under the dose–toxicity curve: $p_1 = 0.10, p_2 = 0.25$, and $p_3 = 0.40$.

Figure 11.3 also shows that the group-of-three initial desi
but longer duration than the group-of-five design; but the di
give a sense about the impacts of the initial design, Figure 1
duration of the TITE-CRM designs under another curve with $p_1 = p_2 = 0.$
$p_{=}0.25$, that is, dose level 3 is the MTD, under two patient accrual rates at $\xi =$
patients per observation window. While we still see that the
slightly superior in terms of PCS and inferior in terms of dur
to patient accrual rate ξ
impact of τ is much more pronounced with a fast accrual rate $\xi = 12$.

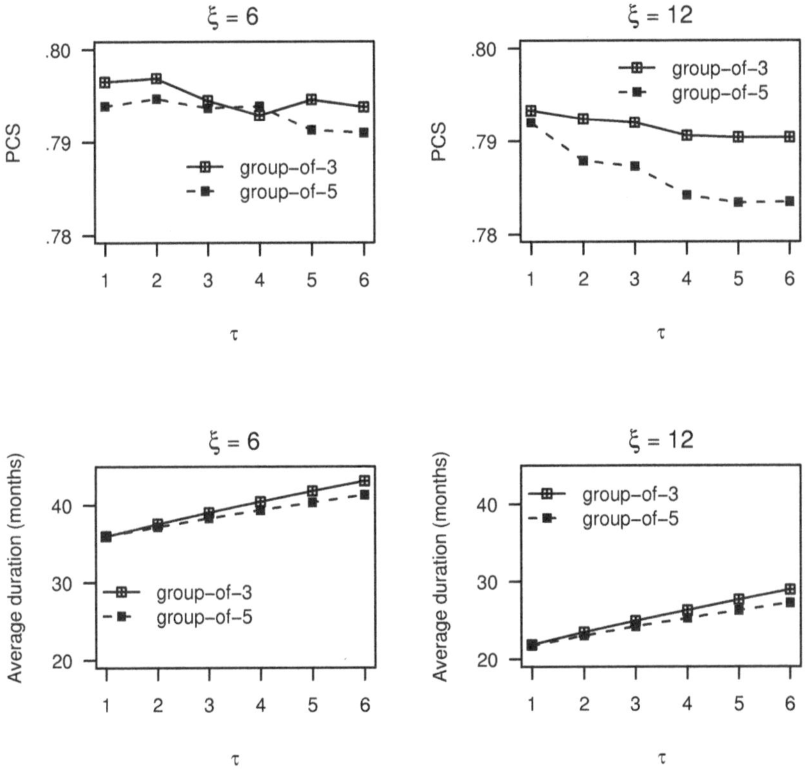

Figure 11.4 Operating characteristics of the two-stage TIT
various initial design under dose–toxicity curve: $p_1 = p_2 = 0.10$ and $p_3 = 0.25$.

11.7 Bibliographic Notes

The problem of long trial duration of the CRM is noted by Goodm
propose the use of group accrual in the CRM. Subsequently, Th

a look-ahead strategy in conjunction with the CRM. The look-
similar steps to Algorithm 11.1 by enumerating all possible
the partially followed patients, and continues accrual onl
recommended doses point to the same dose. Comparatively, du
to the look-ahead strategy is less substantial than the group accrual approach.

The time-to-event formulation of the MTD is first introduced
Chappell [20], who also propose the TITE-CRM. The first docum
trial is published by Muler et al. [69]. Hüsing et al. [47] co
design analogous to the look-ahead extension of the 3+3 algo
times-to-toxicity are incorporated in dose decisions thro
function.

There are several proposals on the weighting schemes direct
safety of the TITE-CRM under late-onset toxicities. Braun [
through estimation under the cure model framework. Cheung [
adjusted adaptive weight in the context of radiation therap
induce late toxicities. Cheung and Thall [24] propose a gene
weights similar to (11.7) and extend the use of the weighted l
clinical trial settings where the binary outcome is defined i
time-to-event variables over a fixed observation window. A r
TITE-CRM geared toward oncologists is given in Normolle and Lawrence [73].

11.8 Exercises and Further Results

Exercise 11.1. A TITE-CRM trial may be generated using the function
the 'dfcrm' package. For example, the following code genera
in Figure 11.2:

```
> ### A simulated trial of two-stage TITE-CRM of 30 subjects
> PI <- c(0.05, 0.10, 0.25)     # true toxicity probabilities
> p0 <- c(0.15, 0.25, 0.35)     # initial guesses
> x0 <- c(rep(1,3),rep(2,3),rep(3,24))  # initial design
> N <- 30
> xi <- 6   # Patient accrual rate
> tau <- 4  # Waiting window
> foo <- titesim(PI,p0,0.25,N,x0,obswin=6,rate=xi,tgrp=tau)
> foo$MTD   # MTD estimate
[1] 3
>
```

The argument `rate` specifies the patient accrual rate ξ
two-stage design, the waiting window τ in stage 1 is specified via `tgrp`
use the same time unit as the observation window `obswin`
accrual patterns are provided by the function via the specifi
`accrual`: setting `accrual="poisson"`
cordance with the Poisson process, whereas `accrual="fixed"`
gives fixed patient enrollment. The other arguments of the fu
those in the function `titecrm`.

Use the function `titesim`
design under various waiting window; cf., Section 11.6.2.

Exercise 11.2. Verify the toxicity risks of the patients given in Table 11.1.

Exercise 11.3 (sketch of proof of Theorem 11.4). Let $t_0 = \min\{t : Y_i(t) = 1\}$
time the first toxic outcome appears.

a. Prove that $\lim_{h \to 0} P_{\mathscr{D}}\{U(t_0, h) > 0\} = 0$ implies Condition (11.12).
b. Assume that $x_{i,0}$ is a nondecreasing sequence. For given $t_0, h >$
$U(t_0, h)$ attains its maximum when $Y_{N(t_0)} = 1$ and $Y_i(t) = 0$ for $i < N(t_0)$
is, the most recent patient experiences toxicity at time t_0,
c. Suppose that $t_0 \in (e_n, e_{n+1}]$
of patients n and $n+1$. Prove that $U(t_0, h)$ attains its maximum when $t_0 = e_n$
d. Based on the above results, show that if an initial sequenc
the regular CRM (i.e., TITE-CRM with $\xi =$
corresponding TITE-CRM regardless of waiting window.

Chapter 12

CRM with Multiparameter Models

12.1 Introduction

We have seen in Chapter 5 that the one-parameter CRM is consis
model misspecifications, but is not generally so (Theorem 5.
due to model misspecification, one may be compelled to consid
dose–toxicity function in the CRM than a one-parameter func
we discuss the use of multiparameter models in the CRM litera
first reviews dose finding methods based on *nonparametric*
curve-free methods; and Section 12.3 presents the potentia
these methods and provides remedies. Section 12.4 explores
CRM and illustrates its pitfalls through numerical example
literature review on the use of these multiparameter models in the CRM.

12.2 Curve-Free Methods

12.2.1 The Basic Approach

The basic idea of curve-free methods is to treat the next grou
with toxicity probability estimated to be closet to θ, that is,

$$x_{i+1} = \arg\min_{d_k} |\hat{\pi}_i(d_k) - \theta|$$

where $\hat{\pi}_i$ is a nonparametric isotonic estimate of the dose–toxicity curve π
the first i groups of observations; that is, the form of $\hat{\pi}_i(x)$
it is monotone increasing. This modeling strategy in effect assumes a K
dose–toxicity curve. Because there is not much borrowing st
across doses in the estimation of the dose–toxicity curve, a
requires treating a few patients at a dose before the next upd
accrual enrollment plan with $m > 1$. Therefore, we let x_i
to the ith *group* of patients in this and the next section. In addition, in alig
tendency for being conservative, these designs usually sta
The main variations among the curve-free methods thus arise
estimation approaches. The following subsection outlines
approaches.

12.2.2 Product-of-Beta Prior Distribution

Gasparini and Eisele [40] propose the use of Bayesian nonpar
dose–toxicity curve. Specifically, the toxicity probabilities p_j
terms of

$$\phi_1 = 1 - p_1 \text{ and } \phi_j = \frac{1 - p_j}{1 - p_{j-1}} \text{ for } j = 2, \ldots, K$$

where each $\phi_j \sim \text{beta}(a_{1j}, a_{2j})$
mean and variance of ϕ_j are, respectively,

$$E(\phi_j) = \frac{a_{1j}}{a_{1j} + a_{2j}} \text{ and } \text{var}(\phi_j) = \frac{a_{1j}a_{2j}}{(a_{1j} + a_{2j})^2 (a_{1j} + a_{2j} + 1)}.$$

Converting (12.2) gives

$$p_k = 1 - \prod_{j=1}^{k} \phi_j$$

so that the beta prior distribution of ϕ_js imposes strict monotonicity, that is, p
$p_2 < \cdots < p_K$, without additional parametric assumptions. The prior on p_k
to follow the product-of-beta distribution [101].

Using (12.3), the binomial likelihood based on i
written as

$$\prod_{k=1}^{K} p_k^{Z_{ki}} (1 - p_k)^{i_k - Z_{ki}} = \prod_{k=1}^{K} \left(1 - \prod_{j=1}^{k} \phi_j\right)^{Z_{ki}} \left(\prod_{j=1}^{k} \phi_j\right)^{i_k - Z_{ki}}$$

where $i_k = i_k(i)$ is the number of patients treated at dose level k
groups of patients, and Z_{ki}
we have $\sum_{k=1}^{K} i_k = im$. Then the $(i+1)$
to (12.1) with

$$\hat{\pi}_i(d_k) = E_i(p_k) = 1 - E_i\left(\prod_{j=1}^{k} \phi_j\right)$$

where $E_i(\cdot)$ denotes expectation taken with respect to the posterior dis
the first i groups of observations.

By the binomial expansion of the likelihood (12.4), the post
ϕ_js can be expressed as a mixture of beta distributions. Thus, a
formula can be derived [40]. However, exact computation wil
unstable as the sample size grows: the computation involves
positive and negative terms, which eventually amount to a va
causes the cancelation of significant digits, a well-known n
one will need to resort to simulation integration methods su
Monte Carlo (MCMC) for general posterior computations [42].

Another practical issue is the elicitation of the hyperparameters (a_{1j}, a_{2j})
product-of-beta prior of ϕ_j. Gasparini and Eisele [40] suggest:

1. Choose the initial guesses p_{0j}s of toxicity probabilities as in the CRM.

2. Treat p_j as if it were a beta variable with hyperparameters (A_{1j}, A_{2j}) determine A_{1j} and A_{2j} by matching the median of beta(A_{1j}, A_{2j}) to p_{0j} that its variance is maximized subject to unimodality. In effect, if $p_{0j} \leq$ then $A_{1j} = \log(0.5)/\log(p_{0j})$ and $A_{2j} = 1$; else, if $p_{0j} > 0.5$, then $A_{1j} =$ $A_{2j} = \log(0.5)/\log(1 - p_{0j})$.
3. Determine the parameters a_{1j} and a_{2j} of the beta prior for ϕ_j two moments via (12.2):

$$E_0(\phi_j) = \frac{a_{1j}}{a_{1j} + a_{2j}} \approx \frac{E_0(1 - p_j)}{E_0(1 - p_{j-1})}$$

$$E_0(\phi_j^2) = \frac{a_{1j}a_{2j}}{(a_{1j} + a_{2j})^2(a_{1j} + a_{2j} + 1)} + \left[\frac{a_{1j}}{a_{1j} + a_{2j}}\right]^2 \approx \frac{E_0(1 - p_j)^2}{E_0(1 - p_{j-1})^2}$$

for $j = 1, \ldots, K$, where $E_0(1 - p_j) = A_{2j}/(A_{1j} + A_{2j})$ and

$$E_0(1 - p_j)^2 = \frac{A_{1j}A_{2j}}{(A_{1j} + A_{2j})^2(A_{1j} + A_{2j} + 1)} + \left[\frac{A_{2j}}{A_{1j} + A_{2j}}\right]^2$$

with $E_0(1 - p_0) = E_0(1 - p_0)^2 = 1$.

According to this algorithm, the prior median of p_j
guess p_{0j}, and mean $E_0(p_j) = A_{1j}/(A_{1j} + A_{2j})$
of the product-of-beta prior in the context of the NeuSTART,
$p_{01} = 0.02, p_{02} = 0.06, p_{03} = 0.10, p_{04} = 0.18$, and $p_{05} = 0.$
that the prior distribution of p_j is skewed to the right for all j
its mean will likely be greater than its median. This explains why $E_0(p_j) > p$
Table 12.1—and such right-skewedness is generally true if $p_{0j} < 0.$
the product-of-beta prior utilizes the initial guesses in a conservative manner.

Table 12.1 A product-of-beta prior for the NeuSTART

j	p_{0j}	(A_{1j}, A_{2j})	(a_{1j}, a_{2j})	$E_0(p_j)$	Corr. between p_j and			
					p_2	p_3	p_4	p_5
1	.02	(0.18, 1)	(1.00, 0.18)	.15	.86	.79	.70	.60
2	.06	(0.25, 1)	(1.18, 0.07)	.20		.92	.81	.70
3	.10	(0.30, 1)	(1.25, 0.06)	.23			.88	.76
4	.18	(0.40, 1)	(1.30, 0.10)	.29				.87
5	.30	(0.58, 1)	(1.40, 0.17)	.37				

While the prior of p_j
that the former is well approximated by the latter. This sugg
computations can be approximated and simplified by assuming $p_j \sim$ beta$(A_{1j}, A$
priori, which is conjugate for the binomial likelihood of p_{0j}
mean can be approximated by

$$E_i(p_j) \approx \frac{A_{1j} + Z_{ji}}{A_{1j} + A_{2j} + i_j}.$$

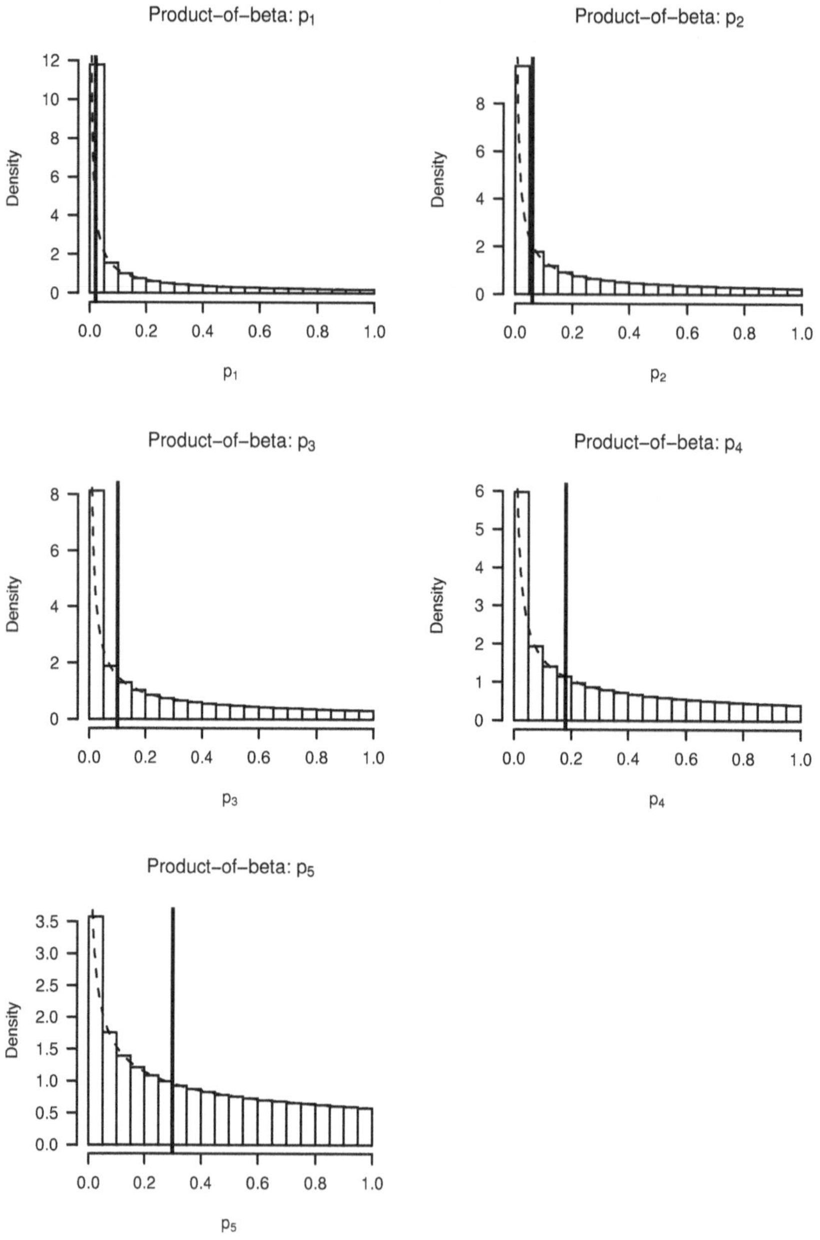

Figure 12.1 Marginal distributions of p_j
In each figure, the dotted line indicates the density of beta(A_{1j},A_{2j})
the initial guess p_{0j} of toxicity probability.

Example 12.1.
NeuSTART, in which the MTD is defined as a dose associated with $\theta = 0.10$. Suppose that the trial starts at dose level 1 and enrolls $m =$ a time, and that the first three patients have no indication of group enters the trial at dose 2, and 1 out of 3 has a toxic outcome. That is, $i_1 = i_2$ $Z_{1i} = 0, Z_{2i} = 1$ with $i =$ of p_1 and p_2 are $E_2(p_1) = 0.11$ and $E_2(p_2) = 0.$ dose level 1. Suppose we use the approximation (12.5), we will get

$$E_2(p_1) \approx 0.04 \text{ and } E_2(p_2) \approx 0.29.$$

which also prescribe dose level 1 for the next group. Suppose three patients is treated at dose 1 with one toxic outcome, so that $i_1 = 6$ and Z_{1i} with $i = 3$. Then the exact posterior means of p_1 and p_2 respective approximation 0.16 and 0.29 according to the beta posterior (12.5).

Example 12.1 illustrates that the beta approximation (12.5 with $i = 3$ and will become increasingly accurate as i thus provides a quick and easy posterior computation instea Exercise 12.2 for the exact computation of $E_i(p_1)$ and $E_i(p_2)$ in this example.

12.2.3 Dirichlet Prior Distribution

An alternative approach to enforce the monotonicity of p_k negative increments of toxicity probabilities. Specifically, let

$$p_k = \sum_{j=1}^{k} \varphi_j$$

for $k = 1, \ldots, K$, where $0 < \varphi_j < 1$ such that $\sum_{j=1}^{K} \varphi_j <$ prior on φ_js is the Dirichlet distribution, that is, having density proportional to

$$\left(\prod_{j=1}^{K} \varphi_j^{c_j - 1} \right) \left(1 - \sum_{j=1}^{K} \varphi_j \right)^{c_{K+1} - 1} \quad \text{for } c_j > 0 \text{ for all } j$$

The likelihood based on i groups of patients can be written as

$$\prod_{k=1}^{K} \left(\sum_{j=1}^{k} \varphi_j \right)^{Z_{ki}} \left(1 - \sum_{j=1}^{k} \varphi_j \right)^{i_k - Z_{ki}}.$$

Using the binomial expansion, we can show that the posterior distribution of a mixture of beta distributions, and derive an exact express of φ_j. However, it has similar numerical problem as in the product and MCMC should be used to obtain posterior quantities as sam moderate.

Under the Dirichlet prior formulation, the prior mean of p_k is $\sum_{j=1}^{k} c_j/C_+$
$C_+ = \sum_{j=1}^{K+1} c_j$ can be viewed as the "prior sample size." We may elicit c_j
the prior mean and the initial guess and solving

$$c_j = (p_{0j} - p_{0,j-1})C_+$$

for $j = 1, \ldots, K$. To determine c_j
prior sample size. To maximize vagueness, one may iterate C_+
integers $\{1, 2, 3, \ldots,\}$ until the prior density of p_k is unimodal for all k
gives the details of the Dirichlet prior using the initial gu
NeuSTART, with $C_+ = \sum_{j=1}^{6} c_j =$
unimodal, and are more informative and have shorter tails th
priors.

Table 12.2 A Dirichlet prior with $C_+ = 2$ for the NeuSTART

j	p_{0j}	c_j	Corr. between p_j and			
			p_2	p_3	p_4	p_5
1	.02	0.04	.56	.42	.30	.21
2	.06	0.08		.75	.53	.38
3	.10	0.08			.71	.51
4	.18	0.16				.71
5	.30	0.24				

12.2.4 Isotonic Design

Leung and Wang [63] take a frequentist nonparametric estima
the toxicity probability of dose level k is estimated using isotonic regression [5]:

$$\tilde{p}_k = \min_{k \le s \le K} \max_{1 \le r \le k} \frac{\sum_{j=r}^{s} Z_{ji}}{\sum_{j=r}^{s} i_j}.$$

The expression (12.6) guarantees that $\tilde{p}_r \le \tilde{p}_k \le \tilde{p}_s$ for $r \le k \le s$
monotonicity assumption. When the sample proportions of to
monotone increasing, that is, $Z_{1i}/i_1 \le Z_{2i}/i_2 \le \cdots \le Z_{Ki}/i_K$,
$\tilde{p}_k$s are equal to the sample proportions.

Similar to the other curve-free methods, the isotonic desig
of patients according to (12.1) with $\hat{\pi}_i(d_k) = \tilde{p}_k$
method, which enforces strict monotonicity through the pri
needed to handle the tied values among $|\tilde{p}_k - \theta|$
regression is highly unsmooth. Suppose the current group is treated at $x_n = d_j$

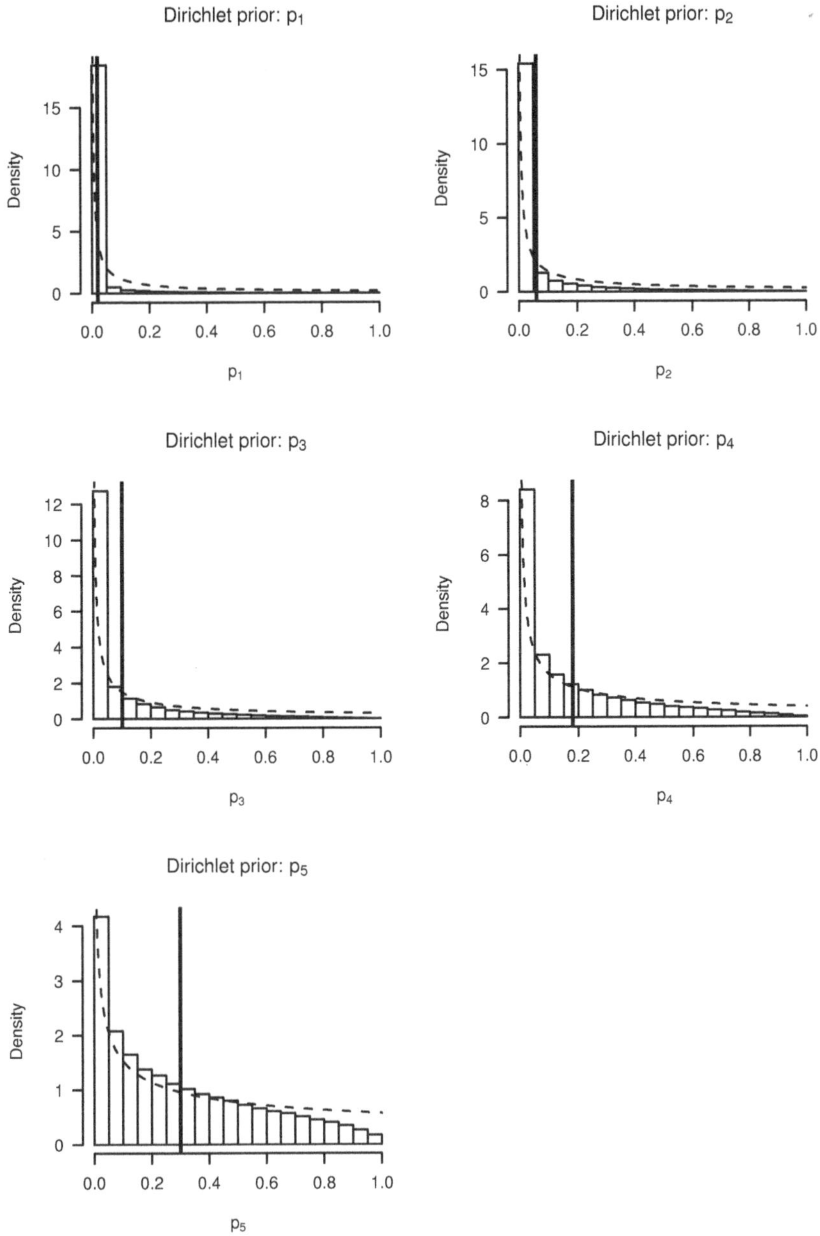

Figure 12.2 Marginal distributions of p_j
each figure, the dotted line indicates the density of beta(A_{1j},A_{2j})
the initial guess p_{0j} of toxicity probability.

the dose for the next is determined as

$$x_{n+1} = \begin{cases} d_{j+1} & \text{if } \tilde{p}_j \le \min(\theta, 2\theta - \tilde{p}_{j+1}) \\ d_{j-1} & \text{if } \tilde{p}_j \ge \theta \text{ and } \tilde{p}_j > 2\theta - \tilde{p}_{j-1} \\ d_j & \text{otherwise.} \end{cases}$$

For example, if $\theta = 0.10$, $\tilde{p}_j = 0.00$, and $\tilde{p}_{j+1} = 0.$
receive dose d_{j+1}.

12.3 Rigidity

12.3.1 Illustrations of the Problem

The conventional intuition for using nonparametric estima
hence enhance the prospect of consistency. However, this co
not hold in dose finding trials where doses assignments are ma
the contrary, the curve-free methods may incur a rigidity pr
assignments to suboptimal levels *as a result of*
approaches:

Definition 12.1 (rigidity). *A dose finding design* $\mathscr{D}$
$0 < p_L < \theta < p_U < 1$ *and all* $i \ge 1$,

$$P_{\mathscr{D}}\{x_i \in \mathscr{I}_\pi(p_L, p_U)\} < 1,$$

where $\mathscr{I}_\pi(p_L, p_U) = \{x : p_L \le \pi(x) \le p_U\}$.

It can be verified that δ
definition of δ-sensitivity. Therefore, the one-parameter CRM, being δ
nonrigid.

In contrast, curve-free methods are rigid. To illustrate th
outcome sequence specified by i_1, Z_{1i}, i_2 for given $Z_{2i} = 1$ and $i_k \equiv 0$ for $k \ge$
the product-of-beta prior, the exact posterior means of p_1 and p_2 are equal to

$$\begin{aligned} E_i(p_1) &= 1 - Q_1\left(1 + \frac{a_{22}}{Q_0}\right) \\ E_i(p_2) &= 1 - Q_1 Q_2\left(1 + \frac{a_{21} + a_{22} + Z_{1i}}{Q_0}\right) \end{aligned}$$

where $Q_0 = (a_{12} + i_2 - 1)(a_{21} + Z_{1i}) + a_{22}(a_{11} + a_{21} + i_1 + i_2 - 1)$,

$$Q_1 = \frac{a_{11} + i_1 + i_2 - Z_{1i} - 1}{a_{11} + a_{21} + i_1 + i_2} \text{ and } Q_2 = \frac{a_{12} + i_2 - 1}{a_{12} + a_{22} + i_2}.$$

See Exercise 12.2. Applying the above expressions with $Z_{1i} = 0$ and $i_2 =$
panel of Figure 12.3 reveals that

$$|E_i(p_1) - 0.1| < |E_i(p_2) - 0.1|$$

for all i_1. In other words, the curve-free method will select dose leve
no toxic outcome (i.e., $Z_{1i} = 0$) in a long string of patients (i.e., as i_1
happens primarily because there is one toxic outcome $Z_{2i} = 1$ among $i_2 =$
at dose level 2.

A similar problem occurs for the isotonic design: Consider t
in Example 12.1. The isotonic estimates for p_1 and p_2 are $\tilde{p}_1 = 0$ and $\tilde{p}_2 = 0.$
the first six patients, and the next group of patients will be t
this point on, the estimate $\tilde{p}_2$ will be at least 0.33, that is, $|\tilde{p}_2 - 0.1| \geq 0.$
the data collected at dose level 1 will not affect estimation of p_2
very toxic). As a result, if $\tilde{p}_1 \leq 0.1$, then $|\tilde{p}_1 - 0.1| < 0.23 \leq |\tilde{p}_2 - 0.1|$
will stay at dose level 1 indefinitely.

These two examples demonstrate that the rigidity problem is
of nonparametric estimation (which does not allow much borr
across doses) in conjunction with a sequential sampling pla
allows the possibility of an "extreme" overestimate of p_2
dose 2; in the above examples, this extremity is reached if th
outcome in a group of three patients. Now suppose that dose 2 i
$p_2 = 0.10$. Then the probability that the trial is confined to the subo
least 0.27, which is nonnegligible and cannot be improved by
size. If the true MTD is among the higher doses, the rigidity p
lower doses will accumulate and impede the method's ability
true MTD.

12.3.2 *Remedy 1: Increase m*

One way to alleviate the rigidity problem is by reducing the p
outcomes by using a larger group size m. For the isotonic design, with $\theta =$
rigidity will occur when there are two or more toxic outcomes in a group of
5 patients. Hence, the rigidity probability at a dose with $p_j = 0.1$ is $\Pr(Z_{ji}$
$i_j = 5) = 0.08$. This remedial measure reduces the rigidity probability
although it does not do away with the problem altogether.

Similarly, for the curve-free CRM, if we enroll patients in groups of size m
with $Z_{1i} = 1$ and $i_2 =$
nontoxic outcomes, that is, $i_1 = 10, Z_{1i} =$

The use of a large m
the trial, and undermine the adaptiveness of a sequential design.

12.3.3 *Remedy 2: Increase Prior Correlations*

When using a Bayesian estimation approach, one may adopt pri
high correlations among p_js so that the estimation of p_j
observations from doses other than d_j.

For the product-of-beta prior, we may introduce strong prio
p_js by reducing the prior variance of ϕ_j for $j = 2, 3, \ldots, K$
use an informative prior distribution. For example, if we re

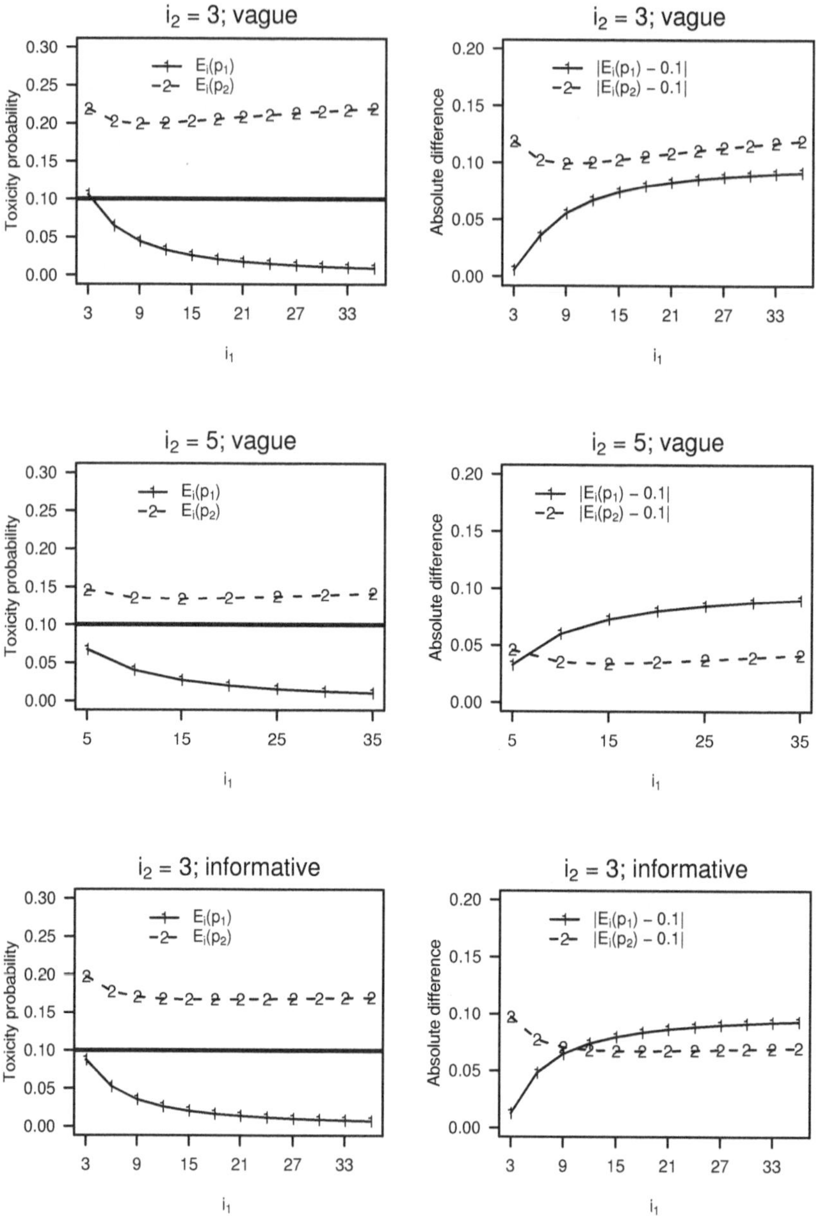

Figure 12.3 Left panel: The posterior means of toxicity prob
1 and 2 against the number i_1 of patients treated at dose level 1 for $Z_{1i} = 0, Z_{2i} = 1$, and i_2
or 5. Right panel: the distance between the posterior means and the target $\theta = 0$.
middle panels: The prior specified in Table 12.1 is used in the
means. Bottom panel: An informative prior in Section 12.3.3 is used.

$a_{12} = 1.00$ and $a_{22} = 0.18$ in Table 12.1 with $a_{12} = 3.53$ and $a_{22} = 0$.
correlation between p_1 and p_2
Figure 12.3 shows that, under this informative prior, the posterior mean $E_i(p$
pulled down to a greater extent by nontoxic observations at d
if we enroll in group size of $m = 3$, with $Z_{1i} = 1$ and $i_2 =$
to dose level 2 after three toxicity-free groups, that is, $i_1 = 9, Z_{1i} = 0$.

Simulation suggests that using an informative product-of-
free method reduces the rigidity probability and shifts the
dose away from the lower ends; and that these are achieved wit
frequency of selecting a toxic dose [15]. It leads to the inve
case where $a_{1j}, a_{2j} \rightarrow \infty$ such that $\phi_j = a_{1j}(a_{1j} + a_{2j})^{-1} \equiv \phi_{0j}$
distribution for $j = 2, \ldots, K$
one-parameter model:

$$p_k \equiv F(d_k, \beta) = 1 - \beta/d_k$$

where the dose labels are $d_1 = 1$ and $d_k = (\phi_{02} \cdots \phi_{0k})^{-1}$ for $k = 2, \ldots, K$, and β is the only free parameter in the model. In addition, if the constants ϕ_{0j}
such that

$$\beta \prod_{j=2}^{k} \phi_{0j} \geq 1 - \theta \text{ for all } k$$

then the limiting model (12.10) will satisfy the conditions
Section 4.3. In light of this, when using the product-of-bet
pointed priors for ϕ_j centering around ϕ_{0j} for $j \geq 2$, where ϕ_{0j}s satisfy (12.11).

The consideration of prior correlations also informs us as t
For example, since the prior correlations under the Dirichl
smaller than those under the product-of-beta prior distrib
appears to be the preferred parameterization [40].

12.4 Two-parameter CRM

12.4.1 The Basic Approach

As the dose finding objective is formulated as a percentile es
shares similar features with the well-studied problem in bi
an extensive set of analytical tools. Detailed expositions
can be found in monographs such as Finney [37], McCullagh and
Morgan [68]. In the literature, it is conventional to use two
capture the shape of a dose–toxicity curve. Some common exam
and the probit models. In particular, the logistic model postulates

$$F(x, \alpha, \beta) = \frac{\exp(\alpha + \beta x)}{1 + \exp(\alpha + \beta x)}$$

where α and β are free parameters to be estimated based on data. Specifical mle $(\tilde{\alpha}_i, \tilde{\beta}_i)$ is obtained by maximizing

$$L_i(\alpha,\beta) = \prod_{j=1}^{i} \left\{ \frac{\exp(\alpha+\beta x_j)}{1+\exp(\alpha+\beta x_j)} \right\}^{Y_j} \left\{ 1 - \frac{\exp(\alpha+\beta x_j)}{1+\exp(\alpha+\beta x_j)} \right\}^{1-Y_j}$$

where x_j and Y_j, respectively, denote the dose and binary outcome of patient j therefore natural to consider using the two-parameter func with the CRM. Also, since a dose finding trial typically confin a discrete set of levels, $\{d_1, \ldots, d_K\}$ observations is given by

$$x_{i+1} = \arg\min_{d_k} \left| F(d_k, \tilde{\alpha}_i, \tilde{\beta}_i) - \theta \right|.$$

Other estimation methods are certainly possible; Chevret [posterior means of α and β is a relatively minor issue. Rather, the main difficulty with arises from the fact that the design points x_i random. This is where the dose finding problem diverges from t problem, and rigidity may occur as a result.

12.4.2 A Rigid Two-Parameter CRM: Illustration

Suppose the outcome sequence in Example 12.1 is observed, th come out of i_1 subjects at dose level 1, and 1 out of 3 at dose level 2. Conside logistic model (12.12) with $x \in \{1, \ldots, K\}$, that is, $d_k \equiv k$ can be simplified as

$$L_i(\alpha,\beta) = \frac{\exp(\alpha+2\beta)}{\{1+\exp(\alpha+\beta)\}^{i_1} \{1+\exp(\alpha+2\beta)\}^3}.$$

The mle $\tilde{\alpha}_i$ and $\tilde{\beta}_i$ can be shown to solve

$$\frac{\partial}{\partial\alpha} \log L_i(\alpha,\beta) = 1 - \frac{i_1 \exp(\alpha+\beta)}{1+\exp(\alpha+\beta)} - \frac{3\exp(\alpha+2\beta)}{1+\exp(\alpha+2\beta)} = 0$$

and

$$\frac{\partial}{\partial\beta} \log L_i(\alpha,\beta) = 2 - \frac{i_1 \exp(\alpha+\beta)}{1+\exp(\alpha+\beta)} - \frac{6\exp(\alpha+2\beta)}{1+\exp(\alpha+2\beta)} = 0$$

from which we can derive

$$F(d_2, \tilde{\alpha}_i, \tilde{\beta}_i) = \frac{\exp(\tilde{\alpha}_i + 2\tilde{\beta}_i)}{1+\exp(\tilde{\alpha}_i + 2\tilde{\beta}_i)} = \frac{1}{3}$$

and

$$\frac{\partial \log L_i(\alpha,\beta)}{\partial\alpha} <$$

at the mle. For technical purposes, suppose the parameters (α, β)
finite rectangle. If α and β
example, the mle does not exist because of (12.17). Then, we c
that

$$F(d_1, \tilde{\alpha}_i, \tilde{\beta}_i) = \frac{\exp(\tilde{\alpha}_i + \tilde{\beta}_i)}{1 + \exp(\tilde{\alpha}_i + \tilde{\beta}_i)} \approx$$

Now, with a target $\theta = 0.$
to dose level 1 as the next recommended dose, that is, $x_{i+1} \equiv d_1$.

To complete the rigidity illustration, we claim that the two
always recommend the lowest dose from this point on, regardl
dose level 1. Exercise 12.3 outlines the proof of this claim.

12.4.3 Three-Stage Design

Like the curve-free method, when the dose–toxicity curve of
is estimated based on very few data points, extreme observat
may confine (with nonnegligible probability) the trial to su
[96] shows that the mles $\tilde{\alpha}_i$ and $\tilde{\beta}_i$
following conditions holds:

$$\begin{array}{ll} (i) & (x^+_{\min}, x^+_{\max}) \cap (x^-_{\min}, x^-_{\max}) \text{ is nonempty} \\ (ii) & x^+_{\min} < x^-_{\min} = x^-_{\max} < x^+_{\max} \\ (iii) & x^-_{\min} < x^+_{\min} = x^+_{\max} < x^-_{\max} \end{array}$$

where $x^+_{\max} = \max\{x_j : Y_j = 1\}$, $x^-_{\max} = \max\{x_j : Y_j = 0\}$, $x^+_{\min} = \min\{x_j : Y_j = 1\}$
$x^-_{\min} = \min\{x_j : Y_j = 0\}$
exist as soon as there is heterogeneity in the toxicity outco
is a weaker condition than (12.19). In other words, it will ta
before the model-based MTD estimate (12.14) can come into ef
one-parameter counterpart (3.11) is used. Thus, it is natur
CRM design if one would like to eventually adopt a two-parameter model:

1. Treat the initial patients according to a predetermined dose sequence $x_{1,0}, \ldots,$
2. Switch to one-parameter CRM once a toxicity is observed among patients.
3. Switch to two-parameter CRM once Condition (12.19) is met.

The stage of using one-parameter CRM may also delay the use of
model until sufficient data are available. Table 12.3a displ
lated trial by a three-stage CRM for the NeuSTART with $\theta = 0.10$ in $N =$
The trial starts with a group-of-three initial design, swit
model used in the NeuSTART (Chapter 10, Section 10.5) upon th
toxic outcome, and uses the two-parameter estimate (12.14)
comes to hold.

The trial is simulated under scenario 3 in Chapter 10, Table 1
level 3 is the true MTD. To summarize, the trial takes 22 patie
parameter model comes into effect, and treats the remaining

$p_4 = 0.0.25$, and $p_5 = 0.30$

Stage	i	u_i	(a) Three-stage two-parameter CRM: x_i	Y_i	Model-based estimates of p_1	p_2	p_3	p_4	p_5	(b) Two-stage one-parameter CRM: x_i	Y_i	Model-based estimates of p_1	p_2	p_3	p_4	p_5
Initial	1	.613	1	0	—	—	—	—	—	Identical to the three-stage CRM						
	2	.636	1	0	—	—	—	—	—							
	3	.780	1	0	—	—	—	—	—							
	4	.058	2	0	—	—	—	—	—							
	5	.189	2	0	—	—	—	—	—							
	6	.018	2	1	.11	.20	.27	.38	.50							
CRM (1p)	7	.992	1	0	.09	.18	.25	.35	.48	Identical to the three-stage CRM						
	8	.891	1	0	.08	.16	.23	.33	.46							
	9	.821	1	0	.07	.15	.21	.32	.44							
	10	.500	1	0	.06	.14	.20	.30	.43							
	11	.165	1	0	.06	.13	.19	.29	.42							
	12	.368	2	0	.05	.12	.17	.27	.40							
	13	.142	2	0	.04	.11	.16	.26	.38							
	14	.645	2	0	.04	.10	.15	.24	.37							
	15	.979	2	0	.04	.09	.14	.23	.36							
	16	.412	2	0	.03	.08	.13	.22	.35							
	17	.905	2	0	.03	.08	.12	.21	.33							
	18	.249	2	0	.03	.07	.12	.20	.33							
	19	.670	3	0	.02	.07	.11	.19	.31							
	20	.499	3	0	.02	.06	.10	.18	.30							
	21	.718	3	0	.02	.06	.09	.17	.29							
CRM (2p)	22	.009	3	1	.02	.07	.28	.66	.90	3	1	.04	.10	.15	.24	.37
	23	.246	2	0	.01	.07	.28	.67	.91	2	0	.04	.09	.14	.23	.36
	24	.727	2	0	.01	.07	.28	.67	.92	2	0	.03	.07	.14	.23	.35
	25	.414	2	0	.01	.06	.27	.68	.92	2	0	.03	.06	.13	.22	.34
	26	.242	2	0	.01	.06	.27	.69	.93	2	0	.03	.06	.13	.21	.34

2, which is the selected MTD in this simulated trial. In fact, with additional patients, the trial will remain at dose 2 or b happens. This example thus illustrates that while rigidity stage design strategy, the problem remains: as the two-para enough to fit well at two dose levels, the estimate for dose 3, $F(d_3, \tilde{\alpha}_i, \tilde{\beta}_i)$ change much after patient 22 and stays close to the sample proportion $\hat{p}_3 = 1/$

A simulated trial by the NeuSTART design is also given in Tabl parison. To match the two designs, the same set of uniform variates $(u_i$ toxicity tolerance is used to generate the outcomes. The one reescalate to dose level 3 after an additional five nontoxic o tients 23 through 27). This example sheds favorable light on when compared to the two-parameter CRM in terms of the rigidi tion, the two-parameter model does not come into play until t through in the example. Since in general, it takes many obser tion (12.19) is satisfied, the typical sample size used in ear may not allow effective use of a two-parameter model.

12.4.4 Continuous Dosage

Piantadosi et al. [81] use a two-parameter CRM in a dose findin continuum of test doses is available. The use of continuous d where using a two-parameter CRM model is reasonable, at leas viewpoint. Since we are allowed to test two doses that are arb other, the rigidity problem is resolved. Ying and Wu [116] pr sequential design, called maximum likelihood recursion, w consistently estimates the target percentile.

However, there are still two practical issues to consider wh model is used. First, a nontrivial number of observations ar Condition (12.19) is satisfied. In the simulated trial in Tab two-thirds of the final sample size before the mles for a two-p To get around with the problem, Piantadosi et al. [81] sugges in model fitting so as to ensure the mle exists. While this appr solution from a computational perspective, the fundamenta informational content in the middle of a trial is so low—few d outcomes—that a two-parameter curve may overfit. Thus, the p should be used with caution.

A second practical issue is that continuous dosage may not be physically plausible in many clinical situations. If a disc trial, we will fall back to the rigidity problem described in will review the stochastic approximation and connect it wit out, the use of discrete test doses is the primary feature tha from the large stochastic approximation literature.

12.5 Bibliographic Notes

Simulation studies in the CRM literature compare the one-pa
with the two-parameter CRM in terms of operating characteri
that the two-parameter CRM does not improve bias and mean squ
compared to the one-parameter CRM. O'Quigley et al. [77] find
CRM could be less efficient than one-parameter CRM; this is tr
two-parameter model used is a correct specification.

In line with the use of underparameterized models in the CRM,
O'Quigley [95] study the general use of one-parameter model
root of a regression function. O'Quigley [74] describes the
associated with fitting a two-parameter model in the CRM.

Gasparini and Eisele [40] and Leung and Wang [63] introduce n
estimation of the dose–toxicity curves in conjunction with
subsequently points out the problem of rigidity with these n

12.6 Exercise and Further Results

Exercise 12.1. Verify the numerical values obtained in Table 12.1 for the sk
$p_{01} = 0.02, p_{02} = 0.06, p_{03} = 0.10, p_{04} = 0.18$, and $p_{05} = 0.30$.

Exercise 12.2. Verify that the exact posterior means in Example 12.1 are giv
(12.8) and (12.9).

Exercise 12.3. Consider the two-parameter CRM used in Section 12.4.2, wher
show that $x_{i+1} = d_1$ after the first six observations given in Example 12.1.

a. Verify that, after the first six observations given, the mles $\tilde{\alpha}_i$ and $\tilde{\beta}_i$ solve

$$Z_{1i} + 1 - \frac{i_1 \exp(\alpha+\beta)}{1+\exp(\alpha+\beta)} - \frac{3\exp(\alpha+2\beta)}{1+\exp(\alpha+2\beta)} = 0$$

and

$$Z_{1i} + 2 - \frac{i_1 \exp(\alpha+\beta)}{1+\exp(\alpha+\beta)} - \frac{6\exp(\alpha+2\beta)}{1+\exp(\alpha+2\beta)} = 0$$

so that (12.16) still holds.

b. Show that $F(d_1, \tilde{\alpha}_i, \tilde{\beta}_i) = Z_{1i}/i_1$ when $Z_{1i} \geq 1$, and hence

$$|F(d_1, \tilde{\alpha}_i, \tilde{\beta}_i) - 0.1| < |F(d_2, \tilde{\alpha}_i, \tilde{\beta}_i) - 0.1|$$

unless $Z_{1i}/i_1 > 1/3$.

The rigidity of the two-parameter CRM can then be completed b
case $Z_{1i} = 0$ using similar arguments as in Section 12.4.2. This exercis
the potential problem that $F(d_1, \tilde{\alpha}_i, \tilde{\beta}_i) > F(d_2, \tilde{\alpha}_i, \tilde{\beta}_i)$
$\tilde{\beta}_i$.

Chapter 13

When the CRM Fails

13.1 Introduction

Most dose finding methods, including the CRM, are suitable fo
satisfy at least one of the following assumptions. First, there is a *strictly*
dose–toxicity relationship around the target dose. Second
an adequate surrogate for efficacy, which often takes much lo
these assumptions are reasonable for traditional cancer ch
batable in other disease areas or treatment options. When ne
the surrogacy perspective of the MTD is not an appropriate op
and as a result the CRM is not a suitable design for dose finding
duces alternative paradigms to the surrogacy perspective o
presents the notion of maximum safe dose, and Section 13.3 ou
ideas of bivariate dose finding in combined phase I/II trials.

13.2 Trade-Off Perspective of MTD

13.2.1 Motivation

The ASCENT trial introduced in Example 2.3 involves six phys
imens with escalating intensity via the timing of the PT sess
session; see Chapter 2, Table 2.1. The target adverse event rate $\theta = 0.$
imately the same as the background adverse event rate in the u
As such, we expect the lower end of the dose–toxicity curve to
as shown in Figure 2.1c. Consequently, the surrogacy definition of the MTD

$$\nu = \arg\min_k |p_k - 0.25|$$

may not be uniquely defined, and dose finding designs that adop
can lead to arbitrary dose selection. Table 13.1 shows the op
a two-stage CRM under three plausible toxicity scenarios fo
On the one hand, the CRM performs in the way it is intended, by selecting *a*
associated with 25% toxicity rate. On the other hand, it is cl
the desired dose is the largest dose with a 25% toxicity rate,
endpoint is not a surrogate for efficacy. (In contrast, if the
work through the toxic side effects, as in cytotoxic agents,

choose any dose with toxicity rate θ
intervals used in Chapter 8.) In other words, one should take
to define the MTD:

$$\gamma = \max\{k : p_k \leq \theta\}$$

where $\theta = 0.25$ for the ASCENT trial.

Table 13.1 Dose selection and the average toxicity number (A
stepwise SPRT procedure for the ASCENT trial

Model	Probability of selecting dose							ATN	ASN[a]
	0	1	2	3	4	5	6		
Scenario L1:	.00	.25	.25	**.25**	.45	.45	.45		
CRM	.00	.47	.20	**.28**	.05	.00	.00	13	50
SPRT	.12	.10	.13	**.60**	.03	.02	.01	16	49
Scenario L2:	.00	.25	.25	.25	**.25**	.45	.45		
CRM	.00	.45	.17	.15	**.20**	.03	.00	13	50
SPRT	.11	.10	.09	.12	**.54**	.03	.01	15	49
Scenario L3:	.00	.25	.25	.25	.25	**.25**	.45		
CRM	.00	.45	.16	.13	.12	**.13**	.02	13	50
SPRT	.12	.09	.09	.08	.11	**.49**	.03	14	47

Note: The CRM uses the logistic model (10.4) specified with th
parameters: $\theta = 0.25$, $K = 6$, $\nu_0 = 3$, $\delta = 0.05$, $N = 50$
from Table 10.3. Numbers associated with the MTD γ *are in bold.*
[a]*ASN = average sample number.*

13.2.2 Maximum Safe Dose and Multiple Testing

A related concept to MTD (13.1) is the maximum safe dose (MAXSD). A dose
said to be safe if $p_k < \vartheta$ for some $\vartheta > \theta$. Then the MAXSD is defined as

$$\text{MAXSD} = \max\{k : p_k < \vartheta\}.$$

In the ASCENT trial, we set $\vartheta = 0.$
at least as large as the MTD. However, the MAXSD (13.2) is equa
under the following class of toxicity scenarios

$$\{\pi : p_k \notin (\theta, \vartheta) \text{ for all } k\}.$$

This class of scenarios is characterized by the exclusion of
zone [7]. Our goal is to estimate the MTD (13.1). However, in l
between the MTD (13.1) and MAXSD (13.2) under (13.3), we can w
objective: design a study to have reasonable estimation pro
under (13.3). Therefore, we set the first design principle fo
as follows:

(P1) Maintaining a reasonably high probability of selectin
all toxicity scenarios belonging to class (13.3).

To avoid additional notation, we shall use γ
unless there is ambiguity. Also, with the notion of an unsafe
to ϑ, we have another obvious design principle:

(P2) Keeping the probability of selecting an unsafe dose low under *all* scenarios.

These two design principles for MTD estimation can be formal
a multiple testing framework. Specifically, consider testi

$$H_{0k} : p_k \geq \vartheta \text{ versus } H_{1k} : p_k < \vartheta \text{ for } k = 1, \ldots, K$$

As in conventional hypothesis testing, a type I error is comm
H_{0k} is falsely rejected, or equivalently, when an unsafe dose is
data. Then the familywise (type I) error rate of a test procedure $\hat{\gamma}$, is defined as

$$\begin{aligned} \text{FWER}(\hat{\gamma}) &\equiv \Pr(\text{Any true } H_{0k} \text{ is rejected}) \\ &= \max_{0 \leq m \leq K} \sup_{\mu \in \Theta_m} P_\pi(\hat{\gamma} > m) \\ &= \Pr(\hat{\gamma} > \text{MAXSD}) \end{aligned}$$

where P_π in (13.5) denotes probability computed under the dose-toxi
and $\Theta_m = \{\pi : p_m < \vartheta, p_k \geq \vartheta; k > m\}$ is the parameter subspace wherein γ
Because of the equality between $\text{FWER}(\hat{\gamma})$
an unsafe dose can be controlled by keeping

$$\text{FWER}(\hat{\gamma}) \leq \alpha_0,$$

an objective to which a statistical test is naturally suited. Since $\{\Theta_m : m = 0, 1$
in (13.5) jointly partition the entire parameter space $[0, 1]^K$
a statistical test) that satisfies (13.7) provide strong con
without any additional assumptions on the parameter space (cf., principle [P2]).

In addition, the selection probability of the true MAXSD (an
under (13.3) can be controlled via the multiple testing framework as follows:

$$\text{PCS}(\hat{\gamma}) = \min_{0 \leq m \leq K} \inf_{\pi \in \Theta_m^*} P_\pi(\hat{\gamma} = m) \geq 1 - \alpha_1$$

where $\Theta_m^* = \{\pi : \max(p_1, \ldots, p_m) \leq \theta, p_k \geq \vartheta; k > m\}$ is a subset of Θ_m
$\cup_{m=0}^{K} \Theta_m^*$ form the scenario class (13.3). Thus, constraint (13.8) in e
probability of selecting the MTD γ
design principle [P1]).

13.2.3 A Sequential Stepwise Procedure

Once the dose finding problem is formulated as a multiple test
number of statistical procedures can, in theory, be used to e

The most well-known procedure is the Bonferroni adjustment
stepwise procedures can be easily adapted to the dose finding
strong control of the FWER [100, 46, 28]. Briefly, a stepwise p
hypothesis from the family (13.4) at a time, and proceeds to t
when certain decisions are reached regarding the current hy
stepwise nature of stepwise procedures, Cheung [18] proposes a two-s
procedure that can be implemented in a *sequential* manner:

Stage 1 An escalation stage where the family (13.4) is tested in the order of

$$H_{01} \rightarrow H_{02} \rightarrow \cdots \rightarrow H_{0K}$$

so that patients are enrolled to dose i only if all doses below i
declared safe, that is, H_{0j} is rejected for all $j < i$. Once a null hypothesis H
accepted for some i, the stepwise procedure will declare dose i
to be unsafe, and it will be unnecessary to enroll patients at
ends upon the first acceptance of a null hypothesis. The seque
and this stopping rule is called a step-down (SD) test. The te
as a mathematical term here, while the sequence (13.9) starts testing a low dose
a high dose. Generally, a SD test starts with testing the most
in this case, H_{01}—and continues to the next restrictive one until the data fai
a null hypothesis.

Stage 2 A deescalation stage where the family (13.4) is tested in the order of

$$H_{0,S_2} \rightarrow H_{0,S_2-1} \rightarrow \cdots \rightarrow H_{01}$$

where S_2 is the dose immediately below the stopping dose in stage 1. Th
are enrolled to dose i only if all doses above i
Deescalation in stage 2 stops once a hypothesis H_{0i} is rejected using the
data accrued in both stages, thus avoid unnecessarily treat
The stopping dose in Stage 2 is used to estimate the MAXSD. The
of testing and this stopping rule is called a step-up (SU) tes
term "step-up" has a mathematical connotation and describe
testing at the least restrictive hypothesis. This two-stag
step-up (SDSU) test [82]. The MAXSD estimate resulted from t
is denoted as $\hat{\gamma}_{SDSU}$.

To make the SDSU procedure more precise, let $\mathscr{R}_{il}$ and $\mathscr{A}_{il}$
the rejection and acceptance regions of H_{0i} based on the *cumulative*
dose i up to stage l for $l = 1, 2$. The starting dose in stage 2 is then given as

$$S_2 = \min\{i : \mathscr{A}_{i1} \text{ is observed}\} - 1$$

and the final MAXSD estimate is given as

$$\hat{\gamma}_{SDSU} = \max\{i : \mathscr{R}_{i1} \cap \mathscr{R}_{i2} \text{ is observed}\}.$$

The operating characteristics of $\hat{\gamma}_{SDSU}$

regions $\mathscr{R}_{il}$ and $\mathscr{A}_{il}$. Let $\xi_i = \Pr(\mathscr{R}_{i1} \mid p_i)$ and $\rho_i = \Pr(\mathscr{R}_{i1} \cap \mathscr{R}_{i2} \mid p_i)$. Then

$$P_\pi(\hat{\gamma}_{SDSU} = m) = \left(\prod_{j=1}^{m-1} \xi_j\right) \rho_m \Omega_{m+1,K}$$

for $m = 1, \ldots, K$, where

$$\Omega_{m,k} = 1 - \xi_m + \sum_{i=m}^{K-1} \left\{ (1 - \xi_{i+1}) \prod_{j=m}^{i} (\xi_j - \rho_j) \right\} + \prod_{j=m}^{K} (\xi_j - \rho_j)$$

with $\Omega_{K+1,K} \equiv 1$. Also,

$$P_\pi(\hat{\gamma}_{SDSU} = 0) = 1 - \sum_{m=1}^{K} P_\pi(\hat{\gamma}_{SDSU} = m).$$

In addition, the key properties $\mathrm{FWER}(\hat{\gamma}_{SDSU})$ and $\mathrm{PCS}(\hat{\gamma}_{SDSU})$
using the expressions (13.12) and (13.13), as follows:

$$\mathrm{FWER}(\hat{\gamma}_{SDSU}) = 1 - P_{\pi_0^L}(\hat{\gamma}_{SDSU} = 0) \text{ and } \mathrm{PCS}(\hat{\gamma}_{SDSU}) = \min_{0 \le m \le K} P_{\pi_m^L}(\hat{\gamma}_{SDSU} =$$

where π_m^L is the so-called *least favorable configuration* under which $p_i = \theta$ for i
and $= \vartheta$ for $i > m$. The simulations in Table 13.1, for example, are done under t
least favorable configurations, $\pi_3^L, \pi_4^L, \pi_5^L$
$\mathscr{R}_{il}$ and $\mathscr{A}_{il}$ that respect the error constraints (13.7) and (13.8) by cali
individual test properties, namely, ξ_i and ρ_i
of $\mathscr{R}_{il}$ and $\mathscr{A}_{i1}$, Cheung [18] details the use of the likelihood ratio test (LR
region that mimics the standard 3+3 algorithm, and the seque
test (SPRT). Furthermore, while (13.12) and (13.13) apply t
that starts at the lowest dose level, the distribution of $\hat{\gamma}_{SDSU}$
is also available [18].

13.2.4 Case Study: The ASCENT Trial

The SDSU procedure with the SPRT, denoted as SDSU-SPRT, was u
ASCENT trial. Let $Z_i(j)$ denote the number of toxic outcomes in the first j
at dose i. The SPRT prescribes test regions

$$\mathscr{R}_{il} = \{\lambda_{i,\tau_{il}} \ge c_l\} \text{ and } \mathscr{A}_{il} = \{\lambda_{i,\tau_{il}} \le c^*\}$$

for $l = 1, 2$, where $\lambda_{i,n}$ is the likelihood ratio for dose i:

$$\lambda_{i,n} = \frac{\theta^{Z_i(n)}(1-\theta)^{n-Z_i(n)}}{\vartheta^{Z_i(n)}(1-\vartheta)^{n-Z_i(n)}}$$

and $\tau_{il} = \inf\{n > 0 : \lambda_{i,n} \ge c_l \text{ or } \le c^*\}$ is a stopping time.

The test regions (13.14) and, hence, SDSU-SPRT are complete

the boundary parameters, $c^* < c_1 < c_2$, and their statistical properties ξ_i and ρ_i well-known approximations in terms of these boundary param SDSU procedure can be calibrated to satisfy the error constr so that

- The probability of erroneous escalation is controlled, that is, $\Pr(\mathscr{R}_{i1} \mid \vartheta) \le \varepsilon^*$
- The average number of patients treated at an overdose is minimized.

See Algorithm 2 in Cheung [18] for details. For the ASCENT tri SDSU-SPRT with the following design parameters and constraints: $K = 6$, $\theta = 0$ $\vartheta = 0.45$, $\alpha_0 = 0.1$, $1 - \alpha_1 = 0.4$, and $\varepsilon^* = 0.5$. and obtained $c^* = 0.24, c_1 = 1.$ and $c_2 = 13.74$. For convenience during the trial, the decision boundari can be charted in terms of the number of toxic outcomes. Table escalation/deescalation decisions according to this SPRT a dose. While the SPRT manages to escalate quickly stage 1 wit nontoxic outcomes, it also is cautious to make a final safety d least nine subjects at a dose (if none exhibits any adverse event).

Table 13.2 Decisions for dose i for the first 20 subjects by the SDSU-SPRT with $c^* = 0$ $c_1 = 1.76$, and $c_2 = 13.74$ in the ASCENT trial

	Stage 1: intervals of $Z_i(j)$			Stage 2: intervals of $Z_i(j)$		
j	Escalate	Stay	Deescalate[a]	Stop[b]	Stay	Deescalate
2	0	[1,2]	—	—	[0,2]	—
3	0	[1,2]	3	—	[0,2]	3
4	0	[1,2]	[3,4]	—	[0,2]	[3,4]
5	[0,1]	[2,3]	[4,5]	—	[0,3]	[4,5]
6	[0,1]	[2,3]	[4,6]	—	[0,3]	[4,6]
7	[0,1]	[2,3]	[4,7]	—	[0,3]	[4,7]
8	[0,2]	[3,4]	[5,8]	—	[0,4]	[5,8]
9	[0,2]	[3,4]	[5,9]	0	[1,4]	[5,9]
10	[0,2]	[3,5]	[6,10]	0	[1,5]	[6,10]
11	[0,3]	[4,5]	[6,11]	0	[1,5]	[6,11]
12	[0,3]	[4,5]	[6,12]	[0,1]	[2,5]	[6,12]
13	[0,3]	[4,6]	[7,13]	[0,1]	[2,6]	[7,13]
14	[0,4]	[5,6]	[7,14]	[0,1]	[2,6]	[7,14]
15	[0,4]	[5,6]	[7,15]	[0,2]	[3,6]	[7,15]
16	[0,4]	[5,7]	[8,16]	[0,2]	[3,7]	[8,16]
17	[0,5]	[6,7]	[8,17]	[0,2]	[3,7]	[8,17]
18	[0,5]	[6,7]	[8,18]	[0,3]	[4,7]	[8,18]
19	[0,5]	[6,8]	[9,19]	[0,3]	[4,8]	[9,19]
20	[0,6]	[7,8]	[9,20]	[0,3]	[4,8]	[9,20]

[a] *Stage 1 ends once a deescalation is invoked.*
[b] *Stage 2 ends and the dose is declared safe once the stopping cr*

The operating characteristics of the ASCENT design under th

configurations are given in Table 13.1, along with the CRM. Th
two theoretical properties of SDSU-SPRT:

- The probability of selecting an overdose is less than $\alpha_0 = 0.10$
- The probability of selecting the MAXSD is at least $1 - \alpha_1 = 0.40$

In fact, the actual performance of SDSU-SPRT is much better t
values α_0 and $1 - \alpha_1$ suggest under these three particular toxicity scenarios. I
that the constraints (13.7) and (13.8) can be quite conserva
for a wide class of dose–toxicity curves.

13.2.5 Practical Notes

While the SDSU-SPRT is tailored for the ASCENT trial where th
curve is flat at the lower tail, it is interesting to see how the
the CRM for situations when the CRM is expected to work. Table
operating characteristics of a two-stage CRM and the ASCENT
configurations where the surrogacy MTD ν
SDSU-SPRT is comparable (or even slightly superior) to the C
selection probability with comparable *average* sample size.

Table 13.3 The operating characteristics of a two-stage CRM
strictly increasing dose–toxicity curves

	0	1	2	3	4	5	6		
Scenario S1	.00	.10	.20	**.25**	.40	.50	.60		
CRM	.00	.03	.32	**.52**	.12	.00	.00	12	50
SPRT	.01	.05	.16	**.65**	.12	.01	.01	17	53
Scenario S2	.00	.05	.05	.18	**.25**	.40	.50		
CRM	.00	.00	.01	.26	**.59**	.14	.00	12	50
SPRT	.00	.00	.04	.16	**.67**	.12	.01	16	54
Scenario S3	.00	.00	.01	.05	.15	**.25**	.45		
CRM	.00	.00	.00	.00	.19	**.70**	.11	11	50
SPRT	.00	.00	.00	.02	.18	**.75**	.04	13	47

*Note: The details of the CRM are given in Table 13.1. Numbers a
MTD are in bold.*

This advantage, however, should be viewed in light of the fac
is an open-ended procedure. Based on Table 13.2, it is possib
more than 20 subjects to a dose in stage 1 if $Z_i(j)$
(defined by the intervals under the Stay column in the table).
small probability for the SPRT to test such a large sample siz
conclusion, the open-endedness of the SPRT renders the proc
infeasible in practice. In contrast, the CRM with a fixed sample size $N =$
with issues involving budget and scheduling.

There are two approaches to prevent an open-ended trial. Fir
statistical test that has with a maximum sample size. For exa
$\mathscr{R}_{il} = \{Z_j(N_l) \leq c_l\}$ where the constants $c_1 \leq c_2$ and $N_1 \leq N_2$
thereby imposing a maximum total sample size KN_2
in Cheung [18], SDSU-LRT is comparable to SDSU-SPRT (and the
of accuracy but will require a slightly larger average sampl
apply a truncated SPRT in the SDSU procedure [114]. This appr
advantage because the SPRT is optimal under the hypothesize
other hand, while truncating a single SPRT is straightforwa
to identify an optimal way to truncate multiple SPRTs with a g
sample size.

In addition to the choice of the target rate θ
specification of an unsafe rate ϑ
for the CRM. In the ASCENT trial, the choice $\theta = 0.$
benefit from a PT regimen is expected to outweigh an increase i
which can be quite severe. The choice of ϑ
explicit control of selection probability under toxicity s
we are in a sense indifferent to whether a dose with $p_i \in (\theta, \vartheta)$
Thus, we may choose ϑ
35% adverse event rate is still acceptable. However, in many
ϑ may be driven by feasibility; generally, the further θ and ϑ
the sample size is needed to satisfy both (13.7) and (13.8). A
we may calibrate the SDSU procedure with respect to the error
ϑ and θ, but will also examine the performance of the procedure unde
narios that contain the indifference zone. For example, alt
for the ASCENT is calibrated with $\vartheta = 0.$
method selects a dose with $p_i = 0.40 \in (0.25, 0.45)$
bility under Scenarios S1 and S2. From a theoretical viewpoi
that the SDSU procedure will select an overdose, that is, $p_i > \theta$, less often as p_i
creases. This property is called *unbiasedness*
nonselection of an overdose even though there is no explicit
of selecting dose i when $p_i \in (\theta, \vartheta)$.

13.3 Bivariate Dose Finding

The bortezomib trial was designed as a phase I/II study, with
solely used toxicity and a testing stage where all patients w
MTD. The ultimate objective was efficacy-driven, with the ho
the identified dose was promising in terms of 2-year progress
but the toxicity-driven dose finding stage was an appropriat
sons. First, toxicity in this trial might be viewed as a surro
the use of a meaningful efficacy endpoint, 2-year progressio
case, would take too long to observe for sequential dose findi
trial, even with a TITE approach (cf. Chapter 11). This secon
finding based on efficacy as infeasible in many situations.

BIVARIATE DOSE FINDING

In some situations, a short-term biomarker may be available as an efficacy endpoint. When this is the case, it is natural to using the efficacy endpoint as well as the toxicity endpoint, whether toxicity is an adequate surrogate. For clarity in cl finding design is defined as a design that uses both toxicity an interim dose decisions. According to this definition, the bo not a bivariate design.

While the CRM in its original prescription is not applicable endpoints, the notion of continual reassessment recurs in t bivariate dose finding; just to name a few, see O'Quigley et al Thall and Cook [102], and Yin et al. [115]. A major theme in mos is the use of model-based and utility-based decisions. To il with a binary efficacy response as well as a binary toxicity outcome. Let $F_E($ and $F_T(x,\beta)$, respectively, denote the model-based probabilities of ac and toxicity at dose x, indexed by a vector-valued parameter β utility $\mathscr{U}(x,\beta)$ of dose x is defined as a function $F_E(x,\beta)$ and $F_T(x,\beta)$ choices of the utility function include

- Probability ratio:
$$\mathscr{U}(x,\beta) = \frac{F_E(x,\beta)}{F_T(x,\beta)};$$
- Odds ratio:
$$\mathscr{U}(x,\beta) = \frac{F_E(x,\beta)\{1-F_T(x,\beta)\}}{\{1-F_E(x,\beta)\}F_T(x,\beta)};$$
- Success probability (under independence of toxicity and efficacy):
$$\mathscr{U}(x,\beta) = F_E(x,\beta)\{1-F_T(x,\beta)\}.$$

The basic idea is to select the dose with the highest utility v practical problem of this utility-based approach is the cho ure 13.1 plots the probability ratio, the odds ratio, and the given dose–toxicity curve and a given dose–efficacy curve. I utility function $\mathscr{U}$ based on the probability ratio or odds ratio utilities, wher when success probability is used. This is hardly surprising (F_E,F_T) on a two-dimensional grid onto a one-dimensional utility $\mathscr{U}$ orderings among some pairs of (F_E,F_T) are crucial in choosing the utility $\mathscr{U}$ relevant.

This being the case, there are useful ideas that may help enha of the trial objective and conduct:

Acceptable doses. sense on the entire dose range. For example, based on the prob Figure 13.1a, the zero dose is the optimal dose. But this is si matical artifact that a virtually zero toxicity probabilit

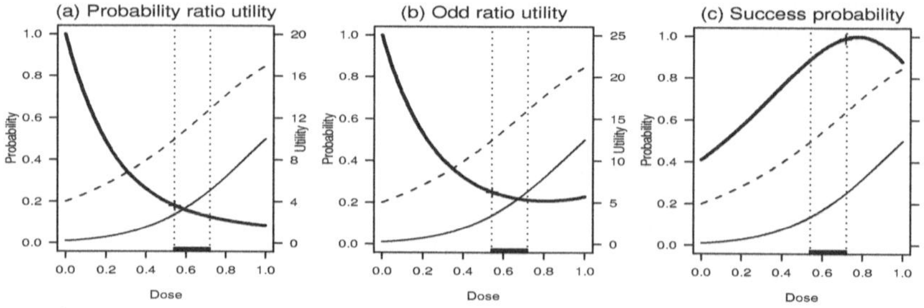

Figure 13.1 Some examples of utility function (dark solid) f
(light solid) and a given dose–efficacy curve (dashed). For a trial with toxicity tolerance θ
0.25 and efficacy threshold $\theta_E = 0.$
region on the x-axis and is bracketed by two vertical dotted l
in each panel indicates the acceptable dose with the largest utility.

its probability ratio. The odds ratio utility also has this p
ure 13.1b. However, based on the success probability in Figu
dose is about 0.80, which is associated with a 30% toxicity ra
considered unacceptable, based on safety. It leads to the no
a dose x is *acceptable* if

$$F_E(x,\beta) \geq \theta_E \text{ and } F_T(x,\beta) \leq \theta_T$$

for some prespecified efficacy threshold θ_E and toxicity tolerance θ_T
view bivariate dose finding as a constrained utility maximiz
(13.15). Bayesian versions of (13.15) have also been adopte

Alternatively, one may set a bivariate dose finding objectiv
an acceptable dose (instead of utility maximization). This ob
than the utility-based approach, and is an easier estimatio
trials where sample sizes are limited, this objective may be
Besides feasibility, this objective also has some clinical
constraints (13.15) are easy to interpret. Furthermore, th
reducing the differences among the utilities. In Figure 13.
range roughly from 0.5 to 0.7: the probability ratio and odds
0.5 on this dose range, and the success probability is maximi
discrepancy among different utility functions is much redu
And importantly, utilities of the acceptable doses have a re
implication is that if we can identify an acceptable dose, we
optimal dose regardless of which utility function we use.

Coherence. The notion of coherence can be easily extended to bivariate d
ing where the toxicity and efficacy endpoints are binary. Wit
comes, there are four possible outcome configurations for each patient:

- Response, toxic

- Nonresponse, toxic
- Response, nontoxic
- Nonresponse, nontoxic

For each of these four possible outcomes, Table 13.4 lists th
are not allowed due to coherence considerations. As discuss
escalation is incoherent after a toxic outcome, and deescal
nontoxic outcome. Also, it seems ethically difficult to give
and, hence, less efficacious dose if the current dose does not
sponse in the current patient. Thus, if the current subject h
toxic outcome, the coherence restrictions will allow neith
tion. It then requires an unambiguous decision that the next
the same dose as the current patient; except that when there i
the current dose is unacceptable with respect to constraint
minate the trial and declare that there is no acceptable dose
in the literature typically involve complex decision algor
this example shows how coherence can simplify clinical deci
defensible.

Table 13.4 Incoherent moves under four possible outcomes

	Toxic outcome	Nontoxic outcome
Response	Incoherent escalation: toxicity	Incoherent deescalation: toxicity
Nonresponse	Incoherent escalation: toxicity Incoherent deescalation: efficacy	Incoherent deescalation: toxicity, efficacy

Our discussion on bivariate dose finding so far assumes we hav
of the dose–toxicity F_T and dose–efficacy F_E
that cannot be taken for granted. As discussed in Chapter 12,
CRM, an elaborate multiparameter model is not necessarily a
it may generate a rigid sequential dose finding design. At the
parameter model (i.e., β is a scalar in both F_E and F_T
to approximate all possible true states of nature. Mimickin
postulated that the model F_E and F_T
sense:

Condition 13.1. *For any given $p_T, p_E \in (0,1)$ and a given x, there exists β parameter space such that $F_E(x,\beta) = p_E$ and $F_T(x,\beta) = p_T$.*

While there is currently no unified modeling theory for bivar
one possible choice is to extend the conditions given in Chap
the bivariate setting and examine their implications. Mode

dose finding (for model-based designs) will be an interestin
to pursue.

Chapter 14

Stochastic Approximation

14.1 Introduction

Robbins and Monro (1951) introduce the first stochastic appr address the problem of finding the root of a regression function $M(x)$ $Y = Y(x)$ denote a random outcome of interest at the stimulus level x $E(Y) = M(x)$. The objective is to sequentially approach the root x^*

$$M(x^*) = \theta$$

for a given θ. In the special case when Y stimulus level is dose, the regression function is equal to t dose x, that is, $M(x) = \pi(x) = \Pr(Y = 1 \mid x)$, and the root x^* is the θ the dose–toxicity curve π to estimating the MTD under the surrogacy perspective (Defin the Robbins-Monro stochastic approximation is a natural me in clinical trials. In this chapter we explore the clinical r approximation method via its connections to the CRM. Sectio Robbins-Monro method and some of its refinements, and discus on the CRM. Section 14.3 presents some adaptations of stocha for dose finding trials. Section 14.4 discuss some future dir methodology. Section 14.5 ends this chapter with some techn the regularity conditions for $M(x)$ and $Y(x)$.

14.2 The Past Literature

14.2.1 The Robbins-Monro Procedure

The Robbins-Monro procedure starts the experiment at a predetermined level x_1 estimates x^* with successive approximation x_i for $i > 1$ recursively:

$$x_{i+1} = x_i - \frac{1}{ib}\{Y(x_i) - \theta\} \quad \text{for some } b >$$

Under mild assumptions about the regression function $M(x)$ and the outcome recursion (14.2) yields a consistent sequence for x^*, that is, $x_i \to x^*$

one. If in addition, the constant b is chosen such that $b < 2M'(x^*)$, then $i^{1/2}(x_i -$
is asymptotically normally distributed with variance

$$\frac{\sigma_0^2}{b\,(2\beta - b)}$$

where $\beta = M'(x^*)$ and $\sigma_0^2 = \text{var}\{Y(x^*)\}$
parametric method in that these convergence results depend
underlying distribution of $Y(x)$ and the regression function $M(x)$
the convergence results are given in Section 14.5.

In practice, the choice of b
for asymptotic normality to hold, the constant b
bound 2β which is unknown. Second, in view of efficiency, the asymptot
(14.3) attains its minimum σ_0^2/β^2 when b is set to β. Since β
situations, these theoretical results are not directly app
can estimate β from the data already observed. Then it is natural to use that
in (14.2) instead of the constant b. To be exact, an *adaptive*
is defined as

$$x_{i+1} = x_i - \frac{1}{ib_i}\{Y(x_i) - \theta\}$$

where $\{b_i\}$ is a sequence of positive random variables such that $b_i \to \beta$
Lai and Robbins [58] study the adaptive procedure (14.4) with b_i
least squares estimate $\hat{\beta}_i$ of β as in a linear model with

$$M(x) = \theta + \beta(x - x^*).$$

Write $Y_j = Y(x_j)$, and let $\bar{x}_i$ and $\bar{Y}_i$ denote the sample means of x_js and Y_j
the first i observations. Then we may set $b_i = \max\{\underline{b}, \min(\hat{\beta}_i, \overline{b})\}$
fied $\underline{b}, \overline{b}$, with

$$\hat{\beta}_i = \frac{\sum_{j=1}^{i}(x_j - \bar{x}_i)(Y_j - \bar{Y}_i)}{\sum_{j=1}^{i}(x_j - \bar{x}_i)^2}.$$

It can be proved that $b_i \to \beta$ and that $\{x_i\}$
asymptotic variance σ_0^2/β^2
the conditions for the general choice of $\{b_i\}$
of the adaptive recursion (14.4).

While the adaptive recursion is theoretically ideal, its im
can be quite limited because estimating the slope β
of data. In dose finding trials with small samples, the additi
least squares estimator $\hat{\beta}_i$ may outweigh its benefits.

14.2.2 Maximum Likelihood Recursion

Suppose that we consider a continuous outcome Y
linear model (14.5), and suppose further that we believe the slope $\beta = b$
natural to estimate x^* with the least squares estimate

$$x_i^* = \bar{x}_i - b^{-1}(\bar{Y}_i - \theta)$$

repeatedly and set the next design point $x_{i+1} = x_i^*$
Robbins [57] show that the design $\{x_i\}$
is identical to the nonadaptive Robbins-Monro method if the same value of b
in (14.2) and (14.7). This identity is quite interesting, as
link between the nonparametric (14.2) and the least squares
parametric model (14.5) with a strong assumption on the slope β
that one may use a parametric model $\tilde{M}(x)$ as a *working model*
sequence, which will be consistent for x^* even though $\tilde{M}(x)$
cation of the true $M(x)$
consistent for x^* if b is not equal to the true β
true $M(x)$ is not linear.

This identity motivates the study of *maximum likelihood recursion*
ing from the exponential family [113]. First, by postulating that $\tilde{M}(x)$
via $x - x^*$, we may rewrite the regression function as $\tilde{M}_c(x - x^*)$ such that $\tilde{M}_c(0$
see the linear model (14.5), for example. Then, the design sequence $\{x_i\}$
recursively based on the maximum likelihood estimate of x^*, that is, obtain x
solving

$$\sum_{j=1}^{i} w(x_j)\,\{Y_j - \tilde{M}_c(x_j - x_i^*)\} =$$

for some appropriately chosen and prespecified weight function $w(x)$, and then

$$x_{i+1} = \max\{x_{\min}, \min(x_i^*, x_{\max})\}$$

where $x_{\min}, x_{\max}$ are, respectively, the lower and upper limits of the dose ran
and Wu [116] show that maximum likelihood recursion (14.9) c
an adaptive stochastic approximation with large samples, s
standard results for stochastic approximation to establis
the maximum likelihood recursion. Briefly, under the regula
in Section 14.5, the recursion formed by (14.8) and (14.9) is consistent for x^*
asymptotically normal. These results hold without assumin
is a correct specification of the true function M.

14.2.3 Implications on the CRM

For binary outcome Y
maximum likelihood recursion, using the logistic function as a working model

$$\tilde{M}(x) = \tilde{M}_c(x - x^*) = \frac{\theta \exp\{b(x - x^*)\}}{1 - \theta + \theta \exp\{b(x - x^*)\}}$$

for some $b > 0$, with $w(x) \equiv$
Ying and Wu [116], we can show that the logit-MLE is asymptoti
is,

$$i^{1/2}(x_i - x^*) \rightarrow_{\mathscr{L}} N\left(0, \frac{1}{b\{2M'(x^*) - b\theta(1-\theta)\}}\right).$$

if $\tilde{M}_c'(0) = b\theta(1-\theta) < 2M'(x^*)$.

If we disregard the truncation (14.9), the logit-MLE effect
subject at the maximum likelihood estimate x_i^*
the likelihood CRM (3.11) if there always exists a test dose d_k with $\tilde{M}(d_k - x_i^*$
Thus, the CRM may be viewed as an analog of the logit-MLE for si
allow a discrete set of test doses. This analogy between thes
some practical insights.

First, the one-parameter CRM is now justified by its (asympto
the nonparametric stochastic approximation via the logit-
parameter location model (14.10). While the theoretical pr
been studied in the recent dose finding literature (cf., Chap
similar model recommendations can be independently reache
the rich literature on stochastic approximation. As a case i
to converge at a $\sqrt{n}$-rate as prescribed in (14.11), the working model $\tilde{M}_c$
relatively shallow, that is, $b < 2M'(x^*)/\{\theta(1-\theta)\}$
the CRM, this suggests that a shallow skeleton $\{p_{0k}\}$
we see that it is asymptotically desirable to choose a workin
CRM has a narrow indifference interval, which indeed corres
of p_{0k}s; see Exercise 8.2. This recommendation thus concurs with t
large sample properties of the logit-MLE.

Second, while the justification of the model-based logit-ML
totic equivalence to the nonparametric Robbins-Monro proc
been demonstrated to be superior to the latter in finite sampl
Thus, a model-based approach seems to retain information th
when using nonparametric procedures. This speculation is m
much confidence about the working model. In a typical dose find
data and small sample sizes, this efficiency comparison lend
model-based methods such as the CRM over the nonparametric a
designs.

14.3 The Present Relevance

14.3.1 Practical Considerations

While the work of Robbins and Monro has motivated numerous re
generated a large number of applications in the statistical
its clinical applications have been rare—for a number of pra
elucidate the three main issues here.

First, as mentioned in the previous section, with a binary ou
samples, the Robbins-Monro method is generally less efficie
methods, and hence may not be suitable for clinical dose findi
endpoint is classified as toxic versus nontoxic.

Second, the method entails the availability of a continuum o
in Chapter 3 in the context of the bortezomib trial and the ASC
unrealistic. In the NeuSTART, where ordering of the test reg
the sole basis of the dose of a drug, dose availability is limi

tablet and convenience of drug preparation. To accommodate that allow only a discrete set of test doses, one may naturall of the stochastic approximation. Suppose that the test doses are labeled {1 A discretized stochastic approximation yields the next dose via

$$x_{i+1} = C\left\{x_i - \frac{1}{ib}(Y_i - \theta)\right\}$$

where $C(x)$ is the rounded value of x for $x \in [0.5, K+0.5)$ K, respectively, if $x < 0.5$ or $\geq K+0.$ rigid (Definition 12.1) and may be stuck at a wrong dose indefin consider applying (14.12) with $b = 0.5$, $\theta = 0.1$, and a starting dose $x_1 =$ be shown that $x_2 = C(1.2) = 1$ if $Y_1 = 0$ and $x_2 = C(-0.8) = 1$ if $Y_1 =$ the second patient will be treated at dose level 1 regardless In the same manner, we can further verify that the trial will i 1. Generally, using simple algebra, we can show that the update x_{i+1} with (14.12) will stay the same as x_i if

$$\left|\frac{1}{ib}(Y_i - \theta)\right| < 0.5.$$

For binary outcomes, Condition (14.13) is implied by $i > 2\max(\theta, 1-\theta)/b$ outcomes of a general variable type, the likelihood that (14.13) occurs is

$$\Pr\left(\left|\frac{1}{ib}(Y_i - \theta)\right| < 0.5\right) \geq 1 - \frac{4E\,(Y_i - \theta)^2}{i^2 b^2}$$

by Chebyshev's inequality. That is, for any given b a finite variance), the discretized recursion (14.12) will q level, which may or may not be the correct dose. This rigidity of rounding the stochastic approximation output to a discre the term $(ib)^{-1}(Y_i - \theta)$ does not contribute to future updates once i large. Hence, stochastic approximation cannot be directly trials with naive discretization.

Third, at small sample sizes, especially in the early enroll squares estimator (14.6) will fluctuate with substantial va negative value or is undefined (when all patients are treated low informational content of a binary outcome aggravates th As a result, the prespecified truncated values $\underline{b}$ and $\overline{b}$ of the trial. This is in essence running a nonadaptive proced b in the recursion (14.2) or $\underline{b}$ and $\overline{b}$ for (14.4) remains a practical issue.

14.3.2 Dichotomized Data

In many situations, the binary toxicity outcome Y observed quantitative biomarker Q, that is, $Y = I\{Q \geq t_0\}$ threshold t_0. The quantitative measurement Q

than Y, and may be used to estimate x^*
consider the regression model

$$Q = R(x) + \sigma_Q(x)e$$

where e is an unobserved noise from a known distribution η
unit variance, so that $E(Q) = R(x)$ and $\text{var}(Q) = \sigma_Q^2(x)$. The mean function
is assumed to satisfy the regularities for the stochastic ap
Section 14.5. Note that model (14.14) allows heteroscedast
different doses. Since biological lab values tend to exhibi
doses, an assumption of homoscedasticity, that is, $\sigma_Q(x) \equiv \sigma_Q$
for practical purposes. Figure 14.1a shows that variabilit
NeuSTART increases with dose as the mean increases. To make t
parsimonious, we further assume that the variability $\sigma_Q(x)$ increases in x
mean function $R(x)$. Apart from these, the forms of $R(x)$ and $\sigma_Q(x)$
Finally, to identify the distribution η
the noise e fits well to a standard normal distribution.

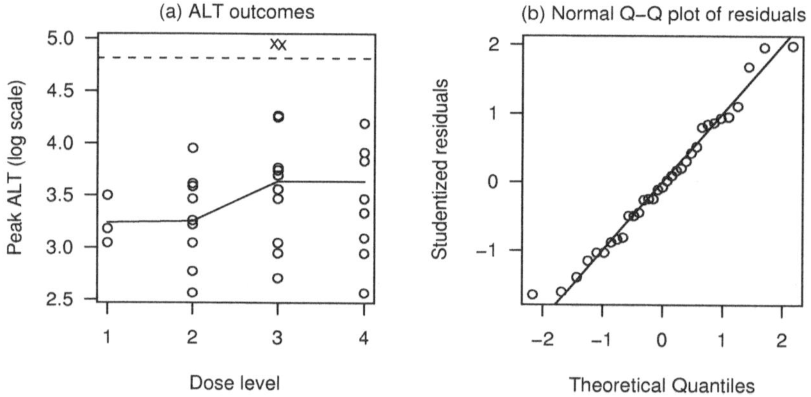

Figure 14.1 (a) The liver enzyme data in the NeuSTART. (b) A di
ity assumption of the noise.

Now, under (14.14), the toxicity probability at dose x
at dose x, can be expressed as

$$M(x) = 1 - \eta\left\{\frac{t_0 - R(x)}{\sigma_Q(x)}\right\}.$$

We can then show that solving the objective (14.1) is equivalent to solving

$$f(x) = t_0$$

where $f(x) = R(x) + c_\theta \sigma_Q(x)$ and c_θ is the upper θth percentile of η.

Suppose for the moment that a continuum of dose x
are enrolled in small groups of size m. Let x_i denote the dose given to the i
and Q_{ij} the measurement of the jth patient in the group. Then define

$$O_i = \bar{Q}_i + \left[E\left\{ \frac{S_i}{\sigma_Q(x_i)} \right\} \right]^{-1} c_\theta S_i$$

where

$$\bar{Q}_i = \frac{1}{m} \sum_{j=1}^{m} Q_{ij} \text{ and } S_i^2 = \frac{1}{m-1} \sum_{j=1}^{m} (Q_{ij} - \bar{Q}_i)^2$$

are, respectively, the sample mean and variance of all measurements in group i
are two important facts about the variable O_i:

- O_i is observable.
- $E(O_i) = f(x_i)$.

See Exercises 14.1 and 14.2 for details. As a result of these t
Robbins-Monro method based on O_is to solve the objective (14.15):

$$x_{i+1} = x_i - \frac{1}{ib}(O_i - t_O).$$

Following standard convergence results, the recursion (14
consistent for x^*, and if in addition $b < 2f'(x^*)$, the sequence $\{x_i\}$
normal with variance

$$v_O = \text{var}\{O_i(x^*)\} \left[b\{2f'(x^*) - b\} \right]^{-1}.$$

In particular, when e is standard normal, for $m \geq 2$,

$$v_O = \frac{\sigma_Q^2(x^*)\{1 + mc_\theta^2(\lambda_m - 1)\}}{mb\{2f'(x^*) - b\}}$$

where

$$\lambda_m = \frac{(m-1)\Gamma^2\{(m-1)/2\}}{2\Gamma^2(m/2)}.$$

and $\Gamma(\cdot)$ is the gamma function.

The efficiency gain due to the use of the quantitative measurement Q
measured by comparing v_O
easily modify the result (14.11) and obtain the asymptotic v
with a group size of $m \geq 2$:

$$v_Y = \frac{1}{mb\{2M'(x^*) - b\theta(1-\theta)\}}.$$

The asymptotic variance v_O is minimized by setting $b = f'(x^*)$
by setting $b = M'(x^*)/\{\theta(1-\theta)\}$ in (14.10). Thus, the optimal choice of b

on the unknown $f'(x^*)$ and $M'(x^*)$
efficiencies, however, suppose we know these optimal values
the variance ratio is equal to

$$\frac{v_Y}{v_O} = \frac{\theta(1-\theta)}{\eta'(c_\theta)\sigma_Q^2(x^*)\{1+mc_\theta^2(\lambda_m-1)\}}$$

where $\eta'(x)$ is the standard normal density. For $m =$
ratio (14.19) is 1.238 when the target $\theta = 0$.
asymptotic relative efficiency of the recursion (14.17), increases as an extreme
targeted [19].

While the efficiency gain due to the use of a quantitative biomarker Q
the use of (14.17) poses two practical issues additional to t
Section 14.3.1. First, the gain is achieved by making an addi
noise distribution η
uses only the dichotomized data Y
of η by looking at pilot data before the trial, and perform diagno
Figure 14.1b). Some preliminary investigation of the impact of a misspecified
given in [22], and further study is warranted.

A second issue is that, in order to be able to estimate $\sigma_Q(x_i)$
enroll patients in groups with size $m \geq$
in a fully sequential manner. This issue does not cause much p
the clinicians are used to the group accrual designs such as the 3+3 algorithm.

14.3.3 Virtual Observations

To address the issue with a discrete design space, Cheung and
the notion of virtual observations. Consider using model (1
notation developed in the previous subsection, we may define
of the ith group of patients as

$$V_i = O_i + b(x_i^* - x_i)$$

where x_i^* denotes the *assigned*
x_i *actually given*
is available for testing), the given dose x_i always equals the assigned dose x_i^*
the far more common situations where $x_i \in \{1,\ldots,K\}$, we may set $x_i = C(x_i^*)$
since the assigned dose x_i^*
the assigned sequence $\{x_i^*\}$
observations V_is:

$$x_{i+1}^* = x_i^* - \frac{1}{ib}(V_i - t_0)$$

To initiate the recursion, we set $x_1^* = x_1 \in \{1,\ldots,K\}$
of starting dose. Note that the given dose x_i
discretized stochastic approximation (14.12). However, t

the assigned dose x^* without rounding. As a result, the term $(ib)^{-1}(V_i - t_0)$ of the order $O(i^{-1})$, can be carried over to the future updates.

Since it is possible that $M(k) \neq \theta$ for all k

$$\nu = \arg\min_k |M(k) - \theta|.$$

Under mild assumptions on $R(x)$ and $\sigma_Q(x)$, the assigned sequence $\{x_i^*\}$ ν_b for some $\nu_b \in \nu \pm 0.5$, and the rounded sequence $\{x_i\}$ to the true ν a dose–toxicity curve where $M(k) \notin (\theta_L, \theta_U)$ for all $k \neq \nu$, for some $\theta_L < \theta <$ consistency requires

$$b < 2\sigma_Q(x^*) \min\left(c_\theta - c_\theta^2/c_L, c_\theta - c_U\right)$$

where $c_L = \eta^{-1}(1-\theta_L)$ and $c_U = \eta^{-1}(1-\theta_U)$ are given in [22].

The idea of virtual observation is to create an objective function

$$h(x) = E(V_i \mid x_i^* = x) = f\{C(x)\} + b\{x - C(x)\}$$

that is defined on the real line, and has a local slope at $\{1, \ldots, K\}$ lution of (14.15) can be approximated by the solution ν_b of $h(x) = t_0$ objective function h has a known slope b around ν_b, we can use the same b cursion (14.21) as in the definition of virtual observations us to achieve asymptotic optimality without resorting to ad slope (cf., Section 14.2). In other words, by using the virtu we also address the practical choice of b of the assigned sequence $\{x_i^*\}$ is inversely proportional to b^2 efficiency, we should choose b permits it. Under the monotonicity assumption of $\sigma_Q(x)$ bound in (14.23) by replacing $\sigma_Q(x^*)$ with an estimate $\hat{\sigma}_Q^*$ for $\sigma_Q(0)$ in untreated patients. Then, we may set b as $2\hat{\sigma}_Q^* \min\left(c_\theta - c_\theta^2/c_L, c_\theta - c_U\right)$.

14.3.4 Quasi-Likelihood Recursion

The concept of virtual observations applies to other outcom instance, the dose finding objective (14.22) is achieved usi Then a virtual observation may be defined as

$$V_i = Y_i + b(x_i^* - x_i)$$

so that $E(V_i \mid x_i^* = x) = M\{C(x)\} + b\{x - C(x)\}$ be generated by stochastic approximation:

$$x_{i+1}^* = x_i^* - (ib)^{-1}(V_i - \theta).$$

However, since the Robbins-Monro method is not efficient wit it is likely to also be the case for the virtual observation re

therefore apply the virtual observation (14.24) with a mode
as the maximum likelihood recursion (14.9). That is, the assigned sequence $\{x_i^*$
generated by recursively solving

$$\sum_{j=1}^{i} w(x_j^*)\{V_j - \tilde{M}_c(x_j^* - x_{i+1}^*)\} =$$

where the model $\tilde{M}_c$
obtained by rounding, that is, $x_{i+1} = C(x_{i+1}^*)$
likelihood estimation [66]. Since the asymptotic properti
recursion rely only on the first two moments of the outcome (cf
expect that the quasi-likelihood recursion (14.26) has sim
under mild assumptions on $\tilde{M}_c$
rigorous manner.

14.4 The Future Challenge

While simulation is an indispensible tool for design evalua
significantly on it), pathological behaviors such as rigidi
be detected in aggregate via simulations. These pathologies are *pointwise*
that can only be found by careful analytical study. In this bo
at a rigorous level, so that the calibration techniques are b
ground. The attempt here, however, primarily addresses the
settings where the objective is to determine the MTD defined i
toxicities over a short-term period. In some chapters, ther
finding in specialized situations such as with late toxiciti
outcomes (Section 13.3), and dichotomized outcomes (Secti
literature has seen an increasing number of designs for thes
a theoretical framework is yet to be worked out before we can s
the calibration of these designs for actual practice.

Obviously, as we deal with different data types other than th
the theoretical criteria presented in this book have to be re
for example. When toxicity is defined by dichotomizing a cont
escalation restrictions should not only take the dichotomi
but also the mean and variability of the continuous observat
suggests, a drug may have a more pronounced impact on the vari
than on its mean level. Therefore, it is sensible to use cauti
increased variability is observed. The formal definition of
yet needs further research.

Despite these complications, it is worthwhile to study the t
dose finding methods in general, especially for the nonstand
coherence again: its extension to the bivariate case in Sect
and clinically sensible guidance on an otherwise complex mo
blackbox. In addition, as the number of parameters used in a b
increases, the concern for rigidity certainly warrants more scrutiny.

How do we proceed from here then? This section illustrates a p through a connection to the rich stochastic approximation l deals with the root-finding objective (14.1) in accordance w definition in the standard dose finding setting. However, the of untapped ideas in the stochastic approximation literatu to dose finding. For example, one may formulate the bivariate problem of finding the maximum of a regression function, for w Kiefer and Wolfowitz [52] and Fabian [35] may prove relevant not covered in this book, individualized dosing is an increa the basic idea is to account for patient heterogeneity by usi covariate [14, 104]. This problem can in fact be reformulate contour-finding problem for which Blum [10] proposes a stoch While it can be an arduous task to study the theoretical prope the CRM to these specialized situations, a connection to the approximation procedures may allow us to do so with relative in the case for the virtual observation recursions (14.21) and (14.26).

14.5 Assumptions on $M(x)$ and $Y(x)$

This section states the conditions on $M(x)$ and $Y(x)$ recursion (14.2) will yield a consistent estimate for the root x^* extracted from Sacks [92] with some slight modifications.

Condition 14.1. *$M(x^*) = \theta$ and $(x - x^*)\{M(x) - \theta\} > 0$ for all $x \neq x^*$.*

Condition 14.2. *$\sup_x var\{Y(x)\} < \infty$ and $\lim_{x \to x^*} var\{Y(x)\}$ exists.*

Condition 14.3. *For some $\varepsilon_1, \varepsilon_2 > 0$,*

$$\sup_{|x - x^*| < \varepsilon_1} E\,|Y(x)|^{2+\varepsilon_2} < \infty.$$

Condition 14.1 ensures that the equation (14.1) has a unique the problem is well defined. Conditions 14.2 and 14.3 impose $Y(x)$ order moments; these two conditions are satisfied when $Y(x)$ Also, we can often modify Condition 14.2 with a slightly stro $\mathrm{var}\{Y(x^*)\} = \sigma_0^2$ for some $\sigma_0 >$ consistency of the Robbins-Monro procedure (14.2), that is, $x_i \to x^*$ one, using the results in Robbins and Siegmund [87].

Condition 14.4. *For some positive constant C, and for all x, $|M(x) - \theta| \leq C|x$*

Condition 14.5. *$M(x) = \alpha_0 + \alpha_1(x - x^*) + \tau(x, x^*)$ such that $\tau(x, \theta) = o(|x -$ as $x \to \theta$.*

Conditions 14.4 and 14.5 impose mild restrictions on the for function $M(x)$. These two conditions, together with Conditions 14.1–14.3 assumption that $b < 2M'(x^*)$, are sufficient for the asymptotic normality of $\{x_i\}$ is,

$$i^{1/2}(x_i - x^*) \to_{\mathscr{L}} N\left(0, \frac{\sigma_0^2}{b\{2M'(x^*) - b\}}\right)$$

where $\sigma_0^2 = \lim_{x \to x^*} \mathrm{var}\{Y(x)\}$
is minimized when the constant b in the recursion (14.2) is set to $M'(x^*)$.

14.6 Exercises and Further Results

Exercise 14.1. Consider regression model (14.14) with a standard normal noise

a. Show that

$$E\left\{\frac{S_i}{\sigma_Q(x_i)}\right\} = \frac{1}{\sqrt{\lambda_m}}$$

where λ_m is defined in (14.18).

b. For the observed outcome O_i defined in (14.16), verify that $E(O_i) = f(x_i)$ show that

$$\mathrm{var}(O_i) = \frac{\sigma_Q^2(x_i)}{m}\{1 + mc_\theta^2(\lambda_m - 1)\}.$$

Exercise 14.2. Repeat Exercise 14.1 for any random noise with $E(e) = \mathrm{var}(e) = 1$. You may consider the following steps:

a. Describe how you will calculate $E\{S_i/\sigma_Q(x_i)\}$ by simulation.

b. Define $\lambda_m = [E\{S_i/\sigma_Q(x_i)\}]^{-2}$. Evaluate λ_m for a logistic noise e and $m = 2$,

c. Verify that $E(O_i) = f(x_i)$ and show that

$$\mathrm{var}(O_i) = \frac{\sigma_Q^2(x_i)}{m}\left\{1 + mc_\theta^2(\lambda_m - 1) + 2mc_\theta\sqrt{\lambda_m}E\left(\bar{e}_i S_i/\sigma_Q(x_i)\right)\right\}.$$

Exercise 14.3. Under model (14.14), show that $M'(x^*) = \eta'(c_\theta)f'(x^*)/\sigma_Q(x^*)$ hence verify the variance ratio (14.19) for normal noise.

References

[1] C. Ahn. An evaluation of phase I cancer clinical trial designs. *Medicine*, 17:1537–1549, 1998.

[2] D. Anbar. Stochastic approximation methods and their us phase I clinical trials. *Communications in Statistics*

[3] M. Ayer, H. D. Brunk, G. M. Ewing, W. T. Reid, and E. Silverm ical distribution function for sampling with incomplete information. *of Mathematical Statistics*, 26:641–647, 1995.

[4] J. Babb, A. Rogatko, and S. Zacks. Cancer phase I clinical dose escalation with overdose control. *Statistics in Medicine* 1998.

[5] D. Bartholomew. Isotonic inference. In *Volume 4*, pages 260–265. New York: Wiley, 1983.

[6] J. Bartroff and T. L. Lai. Approximate dynamic programmi tions to the design of phase I cancer trials. *Statistical Science*

[7] R. E. Bechhofer, J. Kiefer, and M. Sobel. *ing Procedures (with Special Reference to Koopman-Darmois* Chicago: University of Chicago Press, 1968.

[8] B. N. Bekele, Y. Ji, Y. Shen, and P. F. Thall. Monitoring la in phase I trials using predicted risks. *Biostatistics*, 9:442–457, 2008.

[9] J. O. Berger. *Statistical Decision Theory and Bayesian Analysis* Springer-Verlag, 1980.

[10] J. R. Blum. Multidimensional stochastic approximation methods. *Mathematical Statistics*, 25:737–744, 1954.

[11] T. M. Braun. The bivariate continual reassessment meth CRM to phase I trials of two competing outcomes. 23:240–256, 2002.

[12] T. M. Braun. Generalizing the TITE-CRM to adapt for earl toxicities. *Statistics in Medicine*, 25:2071–2083, 2006.

[13] P. P. Carbone, M. J. Krant, S. P. Miller, T. C. Hall, B. I. S J., J. Horton, H. Hosley, J. M. Miller, E. Frie, and M. Schneid feasibility of using randomization schemes early in the cli chemotherapeutic agents: Hydroxyurea. *peutics*, 6:17–24, 1965.

[14] J. D. Cheng, J. S. Babb, C. Langer, S. Aamdal, F. Robert, L
O. Fernberg, J. Schiller, G. Forsberg, R. K. Alpaugh, L. M. We
gatko. Individualized patient dosing in phase I clinical tr
lation With Overdose Control in PNU-214936.
22:602–609, 2004.

[15] Y. K. Cheung. On the use of nonparametric curves in phase
toxicity tolerance. *Biometrics*, 58:237–240, 2002.

[16] Y. K. Cheung. Coherence principles in dose finding studies.
92:863–873, 2005.

[17] Y. K. Cheung. Dose finding with delayed binary outcomes i
S. Chevret, editor, *Statistical Methods for Dose Finding Experiments*
225–242. New York: Wiley, 2006.

[18] Y. K. Cheung. Sequential implementation of stepwise pr
fying the maximum tolerated dose.
ciation, 102:1448–1461, 2007.

[19] Y. K. Cheung. Stochastic approximation and modern mode
dose finding clinical trials. *Statist. Sci.*, 25(2):191–201, 2010.

[20] Y. K. Cheung and R. Chappell. Sequential designs for pha
with late-onset toxicities. *Biometrics*, 56:1177–1182, 2000.

[21] Y. K. Cheung and R. Chappell. A simple technique to evalu
tivity in the continual reassessment method. *Biometrics*

[22] Y. K. Cheung and M. S. V. Elkind. Stochastic approximati
servations for dose finding on discrete levels. *Biometrika*

[23] Y. K. Cheung, P. H. Gordon, and B. Levin. Selecting promi
in clinical trials. *Neurology*, 67:1748–1751, 2006.

[24] Y. K. Cheung and P. F. Thall. Monitoring the rates of comp
censored data in phase II clinical trials. *Biometrics*, 58:89–97, 2002.

[25] S. Chevret. The continual reassessment method in cance
trials: a simulation study. *Statistics in Medicine*, 12:1093–1108, 1993.

[26] S. Chevret, editor. *Statistical Methods for Dose Finding Experiments*
York: Wiley, 2006.

[27] S. Chevret and S. Zohar. The continual reassessment met
editor, *Statistical Methods for Dose Finding Experiments*
New York: Wiley, 2006.

[28] C. W. Dunnett and A. C. Tamhane. A step-up multiple test procedure.
of the American Statistical Association, 87:162–170, 1992.

[29] S. D. Durham and N. Flournoy. Random walks for quantile e
S.S. Gupta and J.O. Berger, editors,
Topics V, pages 467–476. New York: Springer-Verlag, 1994.

[30] S. D. Durham and N. Flournoy. Up-and-down designs I: Sta
distributions. In N. Flournoy and W. F. Rosenberger, editors,

signs, pages 139–157. Hayward, California: Institute of Mathema tics, 1995.

[31] S. D. Durham, N. Flournoy, and A. A. Montazer-Haghighi. designs II: exact treatment moments. In N. Flournoy and W. F. editors, *Adaptive Designs* Mathematical Statistics, 1995.

[32] S. D. Durham, N. Flournoy, and W. F. Rosenberger. A rando phase I clinical trials. *Biometrics*, 53:745–760, 1997.

[33] B. H. Eichhorn and S. Zacks. Sequential search of an optimal dosage, I. *nal of the American Statistical Association*, 68:594–598, 1973.

[34] M. S. Elkind, R. L. Sacco, R. B. Macarthur, D. J. Fink, E. P drews, G. Neils, J. Stillman, J. Chong, S. Connolly, T. Corpo K. Cheung. The neuroprotection with statin therapy for acut (NeuSTART): an adaptive design phase I dose-escalation stu lovastatin in acute ischemic stroke. *International Journal of Stroke* 218, 2008.

[35] V. Fabian. Stochastic approximation of minima with imp speed. *Annals of Mathematical Statistics*, 38:191–200, 1967.

[36] D. Faries. Practical modifications of the continual rea phase I cancer trials. *Journal of Biopharmaceutical Statistics* 1994.

[37] D. J. Finney. *Statistical Methods in Biological Assay*

[38] M. Fisher, K. Cheung, G. Howard, and S. Warach. New pathw ing potential acute stroke therapies. *International Journal of Stroke* 2006.

[39] E. Garrett-Mayer. The continual reassessment method f a tutorial. *Clinical Trials*, 3:57–71, 2006.

[40] M. Gasparini and J. Eisele. A curve-free method for phas *Biometrics*, 56:609–615, 2000.

[41] Constantine Gatsonis and Joel B. Greenhouse. Bayesian clinical trials. *Statistics in Medicine*, 11:1377–1389, 1992.

[42] W. R. Gilks, S. Richardson, and D. J. Spiegelhalter, editors. *Monte Carlo in Practice*. London: Chapman & Hall, 1996.

[43] S. N. Goodman, M. L. Zahurak, and S. Piantadosi. Some pra ments in the continual reassessment method for phase I studies. *Medicine*, 14:1149–1161, 1995.

[44] L. M. Haines, I. Perevozskaya, and W. F. Rosenberger. Ba sign for phase I clinical trials. *Biometrics*, 59:591–600, 2003.

[45] J. M. Heyd and B. P. Carlin. Adaptive design improvement reassessment method for phase I studies. *Statistics in Medicine* 1321, 1998.

[46] J. C. Hsu and R. L. Berger. Stepwise confidence intervals adjustment for dose-response and toxicity studies. *Statistical Association*, 94:468–482, 1999.

[47] J. Hüsing, W. Sauerwein, K. Hideghéty, and K. H. Jöck a dose-escalation study when the event is lagged. 20:3323–3334, 2001.

[48] A. Iasonos, A. S. Wilton, E. R. Riedel, V. E. Seshan, and D comprehensive comparison of the continual reassessment me dard 3+3 dose escalation scheme in phase I dose finding studies. *Trials*, 5:465–477, 2008.

[49] Y. Ji, Y. Li, and B. N. Bekele. Dose finding in phase I clini toxicity probability intervals. *Clinical Trials*, 4:235–244, 2007.

[50] V. R. Joseph. Efficient Robbins-Monro procedure for binary data. 91:461–470, 2004.

[51] H. L. Kaufman, S. Cohen, K. Cheung, G. DeRaffele, J. Mith roziewicz, J. Schlom, and C. Hesdorffer. Local delivery of v expressing multiple costimulatory molecules for the treat tumors. *Human Gene Therapy*, 17:239–244, 2006.

[52] J. Kiefer and J. Wolfowitz. Stochastic estimation of th sion function. *Annals of Mathematical Statistics*, 23:462–466, 1952.

[53] E. L. Korn. Nontoxicity endpoints in phase I trial desig cytotoxic agents. *Journal of National Cancer Institute*

[54] E. L. Korn, D. Midthune, T. T. Chen, L. V. Rubinstein, M. C and R. M. Simon. A comparison of two phase I trial designs. *Medicine*, 13:1799–1806, 1994.

[55] R. Kurzrock and R. S. Benjamin. Risks and benefits of phas revisited. *New England Journal of Medicine*, 352:930–932, 2005.

[56] T. L. Lai. Stochastic approximation. *Annals of Statistics*

[57] T. L. Lai and H. Robbins. Adaptive design and stochastic *Annals of Statistics*, 7:1196–1221, 1979.

[58] T. L. Lai and H. Robbins. Consistency and asymptotic effi estimates in stochastic approximation schemes. *lichkeitstheorie und Verwandte Gebiete*, 56:329–360, 1981.

[59] C. Le Tourneau, J. J. Lee, and L. L. Siu. Dose escalation m cancer clinical trials. *Journal of the National Cancer Institute* 2009.

[60] S. M. Lee and Y. K. Cheung. Model calibration in the conti method. *Clinical Trials*, 6:227–238, 2009.

[61] E. L. Lehmann. *Theory of Point Estimation*. New York: Wiley, 1983.

[62] J. P. Leonard, R. R. Furman, Y. K. Cheung, E. J. Feldman, H Vose, G. Nichols, P. W. Glynn, M. A. Joyce, J. Ketas, J. Ruan, J

R. Niesvizky, A. LaCasce, A. Chadburn, E. Cesarman, and M. Co
I/II trial of bortezomib plus CHOP-Rituximab in diffuse lar
and mantle cell lymphoma (MCL): Phase I results. *Blood*
2005.

[63] D. H. Y. Leung and Y. G. Wang. Isotonic designs for phase I trials. *Clinical Trials*, 22:126–138, 2001.

[64] Y. Lin and W. J. Shih. Statistical properties of the trad
designs for phase I cancer trials. *Biostatistics*, 2:203–215, 2001.

[65] S. M. Lippman, S. E. Benner, and W. K. Hong. Cancer chemop
Journal of Clinical Oncology, 12:851–873, 1994.

[66] P. McCullagh and J. A. Nelder. *Generalized Linear Models*
man & Hall/CRC, 1989.

[67] S. Møller. An extension of the continual reassessment m
liminary up-and-down design in dose finding study in cancer p
to investigate a greater range of doses. *Statistics in Medicine*
1995.

[68] B. J. T. Morgan. *Analysis of Quantal Response Data*
Hall, 1992.

[69] J. H. Muler, C. J. McGinn, D. Normolle, T. Lawrence, D. Br
and M. M. Zalupski. Phase I trial using a time-to-event conti
strategy for dose escalation of cisplatin combined with gem
tion therapy in pancreatic cancer. *Journal of Clinical Oncology*
2004.

[70] J. R. Murphy and D. L. Hall. A logistic dose-ranging meth
ical investigations trials. *Journal of Biopharmaceutical Statistics*
647, 1997.

[71] National Cancer Institute. Common terminology criter
Version 3.0. http://ctep.cancer.gov/forms/CTCAEv3.pdf.

[72] J. Naylor and A. Smith. Applications of a method for the e
of posterior distributions. *Applied Statistics*, 31:214–225, 1982.

[73] D. Normolle and T. Lawrence. Designing dose-escalatio
toxicities using the time-to-event continual reassessment method.
Clinical Oncology, 24:4426–4433, 2006.

[74] J. O'Quigley. Theoretical study of the continual reassessment method.
of Statistical Planning and Inference, 136:1765–1780, 2006.

[75] J. O'Quigley and S. Chevret. Methods for dose finding stu
ical trials: a review and results of a Monte Carlo study.
19:1647–1664, 1991.

[76] J. O'Quigley, M. D. Hughes, and T. Fenton. Dose finding de
studies. *Biometrics*, 57:1018–1029, 2001.

[77] J. O'Quigley, X. Paoletti, and J. MacCario. Non-parame

dose finding studies. *Biostatistics*, 3:51–6, 2002.

[78] J. O'Quigley, M. Pepe, and L. Fisher. Continual reasses
tical design for phase I clinical trials in cancer. *Biometrics*

[79] J. O'Quigley and E. Reiner. A stopping rule for the conti
method. *Biometrika*, 85:741–748, 1998.

[80] J. O'Quigley and L. Z. Shen. Continual reassessment met
approach. *Biometrics*, 52:673–684, 1996.

[81] S. Piantadosi, J. D. Fisher, and S. A. Grossman. Practic
the continual reassessment method for dose finding trials.
apy and Pharmacology, 41:429–436, 1998.

[82] M.-Y. Polley and Y. K. Cheung. Two-stage designs for dos
a biologic endpoint using stepwise tests. *Biometrics*, 64:232–241, 2008.

[83] R Development Core Team.
Computing. R Foundation for Statistical Computing, Vienna, Austria.
3-900051-07-0, URL http://www.r-project.org, 2008.

[84] M. J. Ratain, D. Collyar, B. Kamen, E. Eisenhauer, T. S. L
icz, S. Turner, and J. L. Wade. Critical role of phase I clinic
treatment. *Journal of Clinical Oncology*, 15:853–859, 1997.

[85] M. J. Ratain, R. Mick, R. L. Schilsky, and M. Siegler. Sta
cal issues in the design and conduct of phase I and II clinical
anticancer agents. *Journal of the National Cancer Institute*
1993.

[86] H. Robbins and S. Monro. A stochastic approximation method.
Mathematical Statistics, 22:400–407, 1951.

[87] H. Robbins and D. Siegmund. A convergence theorem for no
most supermartingales and some applications. In Jagdish S.
Optimizing Methods in Statistics
1971.

[88] A. Rogatko, D. Schoeneck, W. Jonas, M. Tighiouart, F. R.
A. Porter. Translation of innovative designs into phase I trials.
Clinical Oncology, 25:4982–4986, 2007.

[89] D. D. Rosa, J. Harris, and G. C. Jayson. The best guess app
trial design. *Journal of Clinical Oncology*, 24:206–208, 2006.

[90] J. Ruan, P. Martin, R. R. Furman, S. M. Lee, K. Cheung, J. M
Casce, J. Morrison, R. Elstrom, S. Ely, A. Chadburn, E. Cesar
man, and J. P. Leonard. Bortezomib plus CHOP-Rituximab for p
untreated diffuse large B cell lymphoma (DLBCL) and mantle c
(MCL). *Journal of Clinical Oncology*, 2010 under review.

[91] D. B. Rubin. Inference and missing data. *Biometrika*, 63:581–592, 1976.

[92] J. Sacks. Asymptotic distribution of stochastic approximation procedures.
nals of Mathematical Statistics, 29:373–405, 1958.

[93] M. A. Schneiderman. How can we find an optimal dose? *Applied Pharmacology*, 7:44–53, 1965.

[94] L. Z. Shen and J. O'Quigley. Consistency of continual re under model misspecification. *Biometrika*, 83:395–405, 1996.

[95] L. Z. Shen and J. O'Quigley. Using a one-parameter model estimate the root of a regression function. *Analysis*, 34:357–369, 2000.

[96] M. J. Silvapulle. On the existence of maximum likelihoo binomial response model. 43:310–313, 1981.

[97] B. Storer and D. DeMets. Current phase I/II designs: Are *Journal of Clinical Research and Drug Development*, 1:121–130, 1987.

[98] B. E. Storer. Design and analysis of phase I clinical trials. *Biometrics* 937, 1989.

[99] M. Stylianou and N. Flournoy. Dose finding using the bias down design and isotonic regression. *Biometrics*, 58:171–177, 2002.

[100] A. C. Tamhane, Y. Hochberg, and C. W. Dunnett. Multiple dose finding. *Biometrics*, 52:21–37, 1996.

[101] J. Tang and A. K. Gupta. On the distribution of the produ beta variables. *Statistics and Probability Letters*, 2:165–168, 1984.

[102] P. F. Thall and J. D. Cook. Dose finding based on efficacy– *Biometrics*, 60:684–693, 2004.

[103] P. F. Thall, J. J. Lee, C. H. Tseng, and E. H. Estey. Accru I trials with delayed patient outcome. *Statistics in Medicine* 1999.

[104] P. F. Thall, H. Nguyen, and E. H. Estey. Patient-specifi bivariate outcomes and covariates. *Biometrics*, 64:1126–1136, 2008.

[105] N. Ting, editor. *and Health)*. Berlin: Springer, 2006.

[106] J. M. Treluyer, S. Zohar, E. Rey, P. Hubert, F. Iserin, M S. Chevret, and G. Pons. A strategy for dose finding and safety based on efficacy and adverse outcomes in phase I/II clinical trials. 54:251–264, 1998.

[107] J. H. Venter. An extension of the Robbins-Monro procedure. *ematical Statistics*, 38:181–190, 1967.

[108] A. Wald. Sequential tests of statistical hypotheses. *Statistics*, 16:117–186, 1945.

[109] A. Wald and J. Wolfowitz. Optimum character of the sequ ratio test. *Annals of Mathematical Statistics*, 19:326–339, 1948.

[110] G. B. Wetherill. Sequential estimation of quantal response curves. *the Royal Statistical Society, Series B*, 25:1–48, 1963.

[111] J. Whitehead and H. Brunier. Bayesian decision proced mining experiments. *Statistics in Medicine*, 14:885–893, 1995.

[112] C. F. J. Wu. Efficient sequential designs with binary data. *American Statistical Association*, 80:974–84, 1985.

[113] C. F. J. Wu. Maximum likelihood recursion and stochast sequential designs. In Van Ryzin, editor, *Related Topics*, volume 8, pages 298–313. Institute of Mathematical Statis 1986.

[114] X. Wu. Stepwise procedures for dose finding in an adapti early rehabilitation after acute stroke. In *DrPH dissertation* versity, 2010.

[115] G. Yin, Y. Li, and Y. Ji. Bayesian dose finding in phase I/ toxicity and efficacy odds ratio. *Biometrics*, 62:777–787, 2006.

[116] Z. Ying and C. F. J. Wu. An asymptotic theory of sequenti maximum likelihood recursion. *Statistica Sinica*, 7:75–91, 1997.

[117] S. Zacks, A. Rogatko, and J. Babb. Optimal bayesian-fe for cancer phase I trials. *Statistics and Probability Letters*

Index